CENTRAL MATHS

FOR GCSE AND STANDARD GRADE

MIKE SMITH AND **IAN JONES**

Heinemann Educational Books

Heinemann Educational Books Ltd
Halley Court, Jordan Hill, Oxford OX2 8EJ

OXFORD LONDON EDINBURGH
MELBOURNE SYDNEY AUCKLAND
IBADAN NAIROBI GABORONE HARARE
KINGSTON PORTSMOUTH NH (USA)
SINGAPORE MADRID

ISBN 0 435 50450 9

Designed by KAG Design Ltd., Basingstoke
Printed and bound in Great Britain by
Scotprint Ltd, Musselburgh

Acknowledgements

The publisher would like to thank the following for their
permission to reproduce photographs or maps:
J Allan Cash (pages 2 (3 photographs of chimney) 6, 9
(Empire State Building, Big Ben, Scott Monument), 10 (Eiffel
Tower), 18, 58 (Edinburgh, York), 68 (radar screen, helicopter),
110, 215, 222, 223 (two road signs), 224 (chair-lift, railway),
229, 231, 234, 244, 246, 268); Allsport/Simon Bruty (page
192); Allsport/Tony Duffy (page 247); Allsport/Duoms
(page 248); Allsport/Vandystadt (page 178); Associated
Sports Photography (page 177); Barnaby's Picture Library
(page 74); Camera Press (page 179); C.E.G.B. (page 220);
Bruce Coleman (page 284); Crown Paints (page 143);
E.H.S. Photography (pages 115, 194, 242, 260 (House of
Commons); Fiat (UK) Ltd (page 170); Financial Times
Library (pages 118 (credit cards), 94 (Neil Kinnock, Margaret
Thatcher)); Gemma Levine/Liberal Party (page 260 (David
Steel)); Ordnance Survey (page 61); Quadrant Picture
Library (page 106); Rank Travel (page 155); Royal Bank of
Scotland (credit card on cover); Samsung (page 269);
Science Photo Library (pages 220 (Gas pipes), 278, 280, 283
(eyelid and blood cell), 285); Brian Shuel (page 53) Sporting
Pictures UK Ltd (page 162); Telefocus (page 10 — Telecom
Tower); Tobler Ltd (page 146); Topham Picture Library
(pages 10 (Blackpool Tower), 238, 278).

Preface

To the student

Maths is a useful and fascinating subject. It enables you to solve problems from all aspects of life. Whether you are buying, renting, making, drawing, designing, building, estimating, or even reading a paper, sooner or later you will be using Maths to solve problems.

We've tried to write a book which includes examples, problems, and investigations from as many aspects of life as possible. We've also included lots of relevant illustrations and photos to make your studies more enjoyable.

Each unit includes an introductory problem from a situation you may be familiar with. You should think about ways of tackling this, but you are not expected to solve the problem straight away. However, by the time you have completed the unit, you should have a good idea how to approach it. If you do have difficulty, a complete solution to the problem is given near the end of the unit, and this is often followed by similar problems for you to tackle.

In this way, we hope that you will enjoy using Maths to solve real problems and that you will learn to use it for problems beyond your studies and this book.

Good Luck.

To the teacher

With the introduction of GCSE and Standard Grade, the emphasis in Maths has moved towards problem-solving and investigations. Coupled with this, there is a need to introduce Maths in contexts which the student sees as relevant to his or her everyday life.

Central Maths has been written with this in mind. It is designed as a core textbook for middle ability students and covers lists 1 and 2 of the National Criteria and the requirements of the various GCSE syllabuses at this level, as well as Standard Grade General Level.

The book consists of 16 units of work, each introduced with a problem to set the Maths in context and promote class discussion. The mathematical skills are presented in short sections of *Information, Examples* and *Exercise*, so that each bank of carefully-graded questions appears alongside the relevant text and worked examples. Once the necessary skills have been covered a solution to the problem is given and further problems are posed. Investigations are suggested throughout the book and provide a possible basis for extended class work.

The order in which the units are taught is in general, unimportant, and teachers can easily design their own course around this book. *Central Maths* can be used as a class resource, for group work, for homework or as a reference book. A full index, listing individual topics, will help students revise before their exams.

It is assumed throughout that students have access to calculators. They are encouraged to estimate each answer before using a calculator, to be critical of the answer, and to consider the degree of accuracy required. Practice in this is given in some of the revision exercises. Unless otherwise stated, answers are given to 3 significant figures.

We hope this book lives up to expectations. If not, alas . . .

Smith and Jones
April 1987

Contents

1 Surveying

Do you remember? ☞ In this unit it is assumed that you know how to:

1   change metric measurements from one unit to another

2   use a protractor.

Exercise 0 will help you check. Can you answer all the questions?

Exercise 0   **Metric measurement**

1   Copy and complete: 1 cm = ☐ mm
                        1 m = ☐ cm
                        1 km = ☐ m

2   Rewrite these measurements in centimetres.

(a) 70 mm   (b) 20 mm   (c) 35 mm   (d) 47 mm

(e) 2 m   (f) 6 m   (g) 4.5 m   (h) 3.75 m

3   Rewrite these measurements in millimetres.

(a) 3 cm   (b) 10 cm   (c) 8.5 cm   (d) 4.2 cm

(e) 50 cm   (f) 60 cm   (g) 1 m   (h) 1.1 m

Using a protractor

4   Use a protractor to measure the size of these angles.

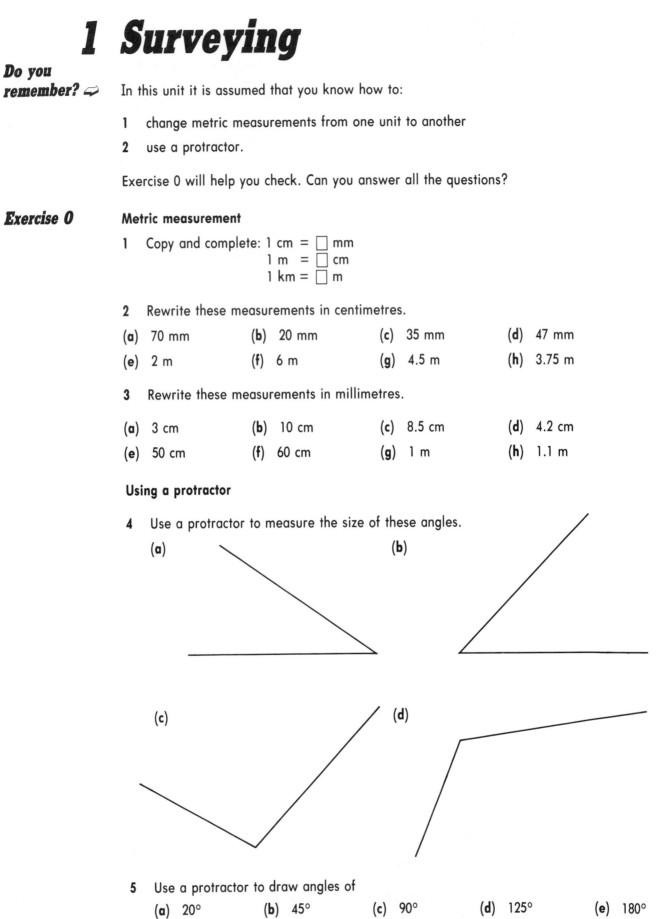

(a)   (b)

(c)   (d)

5   Use a protractor to draw angles of

(a) 20°   (b) 45°   (c) 90°   (d) 125°   (e) 180°

A tall story

A tall chimney, which is close to a small housing estate, is to be demolished. The easiest way to do this is to place explosives at the base of the chimney.

Safety regulations state that this can only be done if there is no building within 30 metres of where the chimney is expected to fall. So if the chimney were 100 metres tall, the nearest building should be at least 130 metres away.

The precise height of the chimney is not known, and it is too dangerous to climb. However, the house nearest to the chimney is found to be exactly 120 metres away.

How would you find out if the chimney could be demolished safely?

120 m

This unit will help you to solve problems like this. It also explains about scale drawings and angles.

Scale drawing

Information ⇨ Many problems can be solved by using **scale drawings**. Measurements from scale drawings are used to calculate corresponding full size measurements. To change measurements from the scale drawing to full size you multiply by the **scale factor**.

Example

Scale drawing to full size

An architect is designing a suspension bridge to take cars over a river to a new shopping centre which is being built. She has made a scale drawing of her design. From her scale drawing work out the height of each tower above the road

Scale 1:1000 (1 cm represents 10 m)

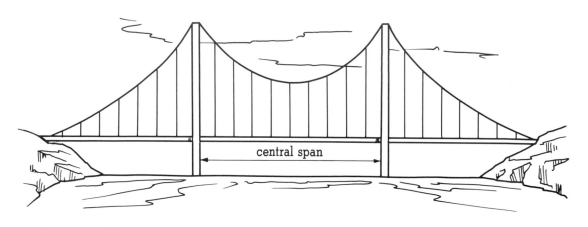

central span

The scale of the drawing is 1:1000. The scale factor is 1000.
To change measurements from the scale drawing to measurements on the real bridge multiply by 1000.
In the scale drawing the height of the tower is 3 cm.

Height of tower on full-size bridge: 3 cm × 1000 = 3000 cm
3000 cm = 30 m

Exercise 1

1 Copy and complete this table. Measure each line accurately from the scale drawing of the bridge.

	Scale drawing	Actual bridge
Height of towers above road Width of river Distance between towers Height of road above water Height of the towers above water	3 cm	3 cm × 1000 = 30 m

2 Calculate the length of each of the 23 vertical cables on the actual bridge.

3 Estimate the total length of the suspension cable on the central span

4 Estimate the total length of the suspension cable to the left of the central span

5 What is the total length of suspension cable needed for the bridge?

6 The cross-section below shows the construction of a child's toy called a rocking duck.

Each side is cut from a single piece of 25 mm thick board. The seat and foot rest are also made from 25 mm thick board.

The dots show the position of the dowels which hold the toy together.

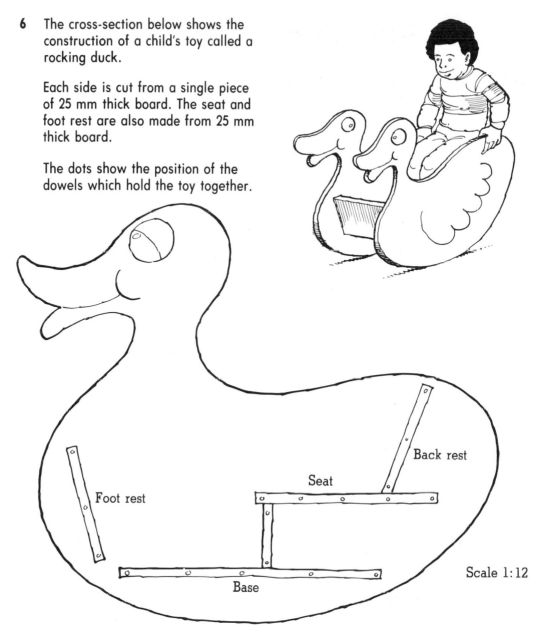

Foot rest

Seat

Back rest

Base

Scale 1:12

(a) How long is the foot rest in the full size toy?

(b) How long is the base in the full size toy?

(c) How long is the seat in the full size toy?

(d) How long is the back rest in the full size toy?

(e) The sides of the toy are each cut from a separate rectangular sheet of board. What size of board is needed for each side?

7 A plan of a car is drawn to a scale of 1:20. The length of the car on the plan is 140 mm. How long is the actual car?

8 A map has a scale of 1:20 000. On the map a path is 8.4 cm long. How long is the path? Give your answer in kilometres and metres.

9 A model aircraft is made to a scale of 1:80. Each wing on the model is 9 cm long. How long are the wings of the real aeroplane?

Information ✍ It is very useful to be able to make scale drawings of full-size objects. You need to do this if you plan layouts or make changes to existing designs. To do this, full size measurements are scaled down. To change measurements from full size to scale size you divide by the scale factor.

Example   **Full size to scale size**

Here is a rough sketch of the garden at the back of a house. *It is not to scale.* What is the length of the garden? What will it be in a scale drawing, drawn to a scale of 1:250?

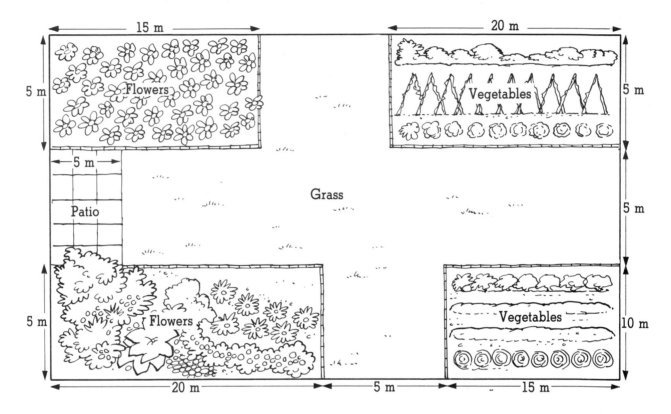

The length of the garden is 40 m. The scale factor is 250.
To change from full size to scale size you must divide.

$$40 \text{ m} \div 250 = 0.16 \text{ m}$$
$$0.16 \text{ m} = 16 \text{ cm}$$

The length of the garden in the scale drawing will be 16 cm.

It is often easier to change the units of measurement first.

$$40 \text{ m} = 4000 \text{ cm}$$
$$4000 \text{ cm} \div 250 = 16 \text{ cm}$$

Exercise 2   1   Make an accurate scale drawing of the garden. Remember to state the scale and give the drawing a title.

2   A new building is designed to be 50 m long. A plan is drawn to a scale of 1:200. How long is the building on the plan?

3   Two villages are 8.5 km apart. How far apart are they on a map drawn to a scale of 1:25 000?

Example

Scaling up and down

The plan of a new double-decker bus is being drawn to a scale of 1:50. If the total length of the bus is 8 m, how long will the plan be ?

You are changing from full size to scale size so you divide.

$$8 \text{ m} = 800 \text{ cm}$$
$$800 \text{ cm} \div 50 = 16 \text{ cm}$$

On the plan a wheel has a diameter of 15 mm. What is the diameter of the full size wheel?

You are changing from scale size to full size so you multiply.

$$15 \text{ mm} \times 50 = 750 \text{ mm}$$
$$750 \text{ mm} = 75 \text{ cm}$$

Exercise 3

1 (a) A new car is 4 m long and 1.5 m high. A scale drawing is to be made to a scale of 1:25. What will be the length and height of the scale drawing?

(b) In the scale drawing the diameter of each wheel is 16 mm. What is the diameter of each car wheel in metres?

2 A scale model of a ship is 13.4 cm long. If the scale of the model is 1:500, what is the length of the ship?

3 On a map drawn to a scale of 1:25 000, the distance betweeen 2 hilltops is 18 cm. How far is it from one hilltop to the other?

4 A classroom measures 8 m long by 5.6 m broad. It is being drawn to a scale of 1:50. What are the length and breadth of the scale drawing?

5 A plan of a building is drawn to a scale of 1:200. On the plan a room measures 95 mm by 125 mm. What are the full size measurements of the room?

6 A map is to be drawn covering a square area 40 km by 40 km. The scale of the map is 1:50 000. What size of paper will be needed to print the map? (Allow a border of 10 cm round each edge.)

7 A classroom measures 11 m by 8 m. A scale drawing of it is to be made on a sheet of A4 paper. Choose a suitable scale.

Angle of elevation

Information ☞ An angle measured from the horizontal in an upward direction is called an **angle of elevation**. An angle measured from the horizontal in a downward direction is called an **angle of depression**. A **clinometer** is a simple device for measuring these angles. The next page shows how you can make a clinometer.

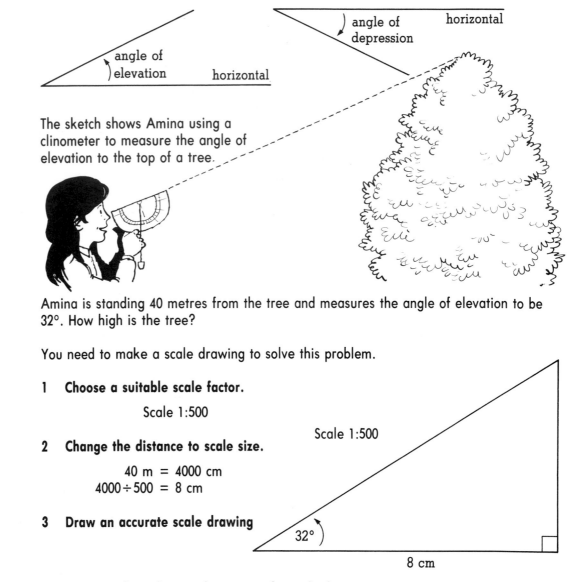

The sketch shows Amina using a clinometer to measure the angle of elevation to the top of a tree.

Example Amina is standing 40 metres from the tree and measures the angle of elevation to be 32°. How high is the tree?

You need to make a scale drawing to solve this problem.

1   Choose a suitable scale factor.

Scale 1:500

2   Change the distance to scale size.

40 m = 4000 cm
4000 ÷ 500 = 8 cm

3   Draw an accurate scale drawing

Scale 1:500

32°

8 cm

4   Measure the unknown distance in the scale drawing.

The vertical side of the right-angled triangle represents the height of the tree. Measure it from the scale drawing. It is 5 cm long.

5   Change this measurement to full size.

5 cm × 500 = 2500 cm
2500 cm = 25 m

Amina is not holding the clinometer at ground level. She is using the clinometer at eye level and you have to take this into account by adding on 1 metre.

The tree is 26 metres tall.

Information ⇨   The following instructions explain how to make a simple clinometer.

You will need a rectangular piece of card measuring 12 cm by 6 cm, a pencil, a ruler, a pair of scissors, a pair of compasses, a protractor, fine string (8 cm long), a small weight.

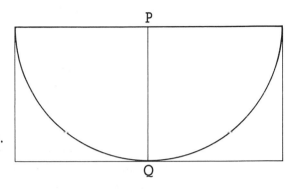

1   Mark the centre points of the top and bottom edges (P and Q). Join them up with a straight line.

2   Using a pair of compasses, draw a semi-circle centre P and radius 6 cm.

3   Cut round the semi-circle. Throw away the 2 corner pieces.

4   Use a protractor to mark off every 5° from the line PQ. Use a longer line for each 10° division.

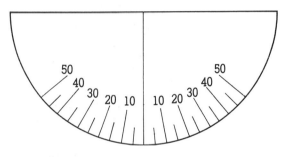

5   Make a plumbline by tying a small weight to the end of the piece of string. Tie a knot in the other end.

6   Cut a notch at P and fit the plumbline in with the knot at P. Make sure that the weight hangs clear of the card.

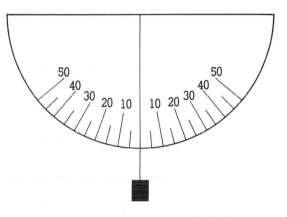

Investigation   Your teacher will give you a list of things round the school, of which to find the height. Here are some hints to help you.

1   Work in pairs.

2   Estimate the height before making any measurements.

3   For good results you should not stand too close to the object.

4   Estimate the object's height and stand twice this distance away.

5   Choose a suitable scale for your drawing.

6   Remember to add on the distance from the ground to your eye level.

Exercise 4

Make scale drawings to find the height of the following structures.
The sketches are not drawn to scale but, to help you, the first six questions have a suggested scale. Answers should be given to the nearest 5 m. The angles of elevation are measured from ground level.

1   Empire State Building, New York.

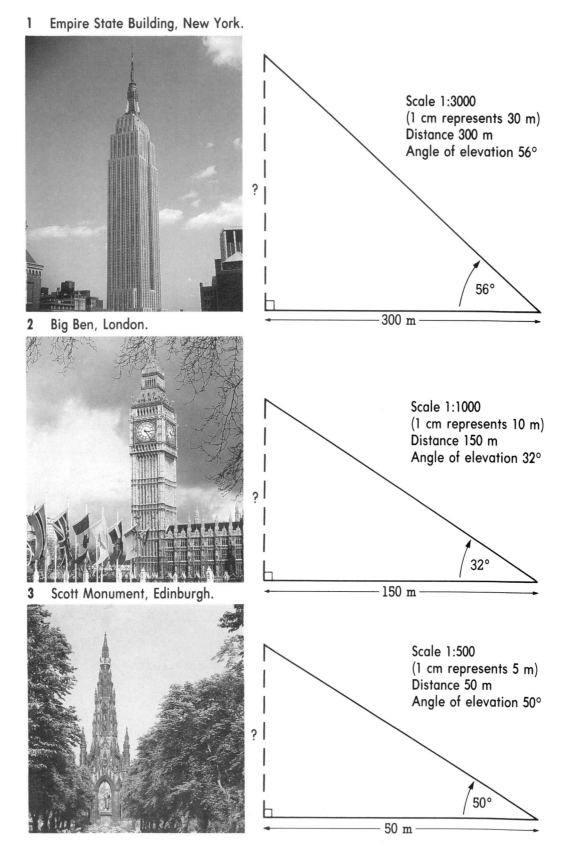

Scale 1:3000
(1 cm represents 30 m)
Distance 300 m
Angle of elevation 56°

300 m

56°

2   Big Ben, London.

Scale 1:1000
(1 cm represents 10 m)
Distance 150 m
Angle of elevation 32°

150 m

32°

3   Scott Monument, Edinburgh.

Scale 1:500
(1 cm represents 5 m)
Distance 50 m
Angle of elevation 50°

50 m

50°

4 Eiffel Tower, Paris.

Scale 1:4000
(1 cm represents 40 m)
Distance 400 m
Angle of elevation 37°

?

?

37°

◄— 400 m —►

5 At a distance of 150 m from the foot of the Telecom Tower, the angle of elevation to the top is 47°. Find the height of the tower.
Suggested scale 1:1500.

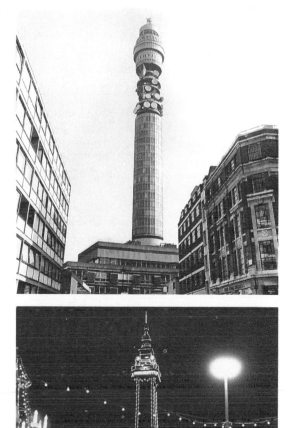

6 The largest trees in the world are Giant Redwoods. Find the height of the tallest Redwood if the angle of elevation is 43° measured 120 m from the base of the tree.
Suggested scale 1:1000.

7 From a distance of 200 m the angle of elevation to the top of the Blackpool Tower is 38°. How high is the tower?

8 In a boat 60 m from the Forth Rail Bridge the angle of elevation to the top was measured to be 61°. How high is the bridge?

9 Salisbury Cathedral is the tallest cathedral in Great Britain. From a distance of 50 m the angle of elevation is 68°. How tall is Salisbury Cathedral.

Scale problems

Information ⇨ Scale drawings can be used to solve many problems. Each of the problems in the exercise below can be solved by making an accurate scale drawing.

Exercise 5

1   At an Army demonstration a death slide is set up from a castle as shown. The starting point is 40 m above ground and the rope is set at an angle of 35° to the horizontal.

(a)   How long is the slide?

(b)   How far away from the castle is the rope anchored?

2   The sketch shows a lean-to greenhouse built onto the side of a house. Find

(a)   the length of glass needed for the sloping roof

(b)   the angle at which the roof slopes to the horizontal.

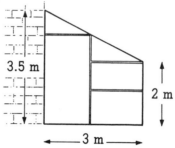

3   Two spotlights above a stage are 8 m apart and angled to light up the same area of stage. One spotlight is angled down at 60° from the horizontal and the other is angled at 35° from the horizontal. How far above the stage are the spotlights?

4   The sketch below, not drawn to scale, shows a section of a swimming pool, 50 m long. The depth of water is to be measured every 10 m. Copy and complete the table below. You will need to make an accurate scale drawing to find the missing depths.

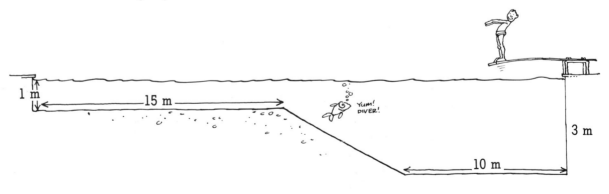

Distance from shallow end (m)	0	10	20	30	40	50
Depth of water (m)	1					3

Is there anything you could do to your drawing so that your answers are more accurate?

Investigation

Plan a Kitchen

Here are the sizes and prices of *Supaline* Kitchen Units.

All units are 870 mm high and 600 mm deep.

Double base unit
1000 mm wide
Price £59.59

Base unit
500 mm or 600 mm wide
£53.19 or £55.59

Single base unit
300 mm wide
£48.39

Corner unit
1000 mm wide
Price £57.19

Larder unit
500 mm wide
£95.39

Drawer unit
500 mm wide
£78.79

A washing machine takes up a width of approximately 600 mm, a fridge 550 mm and a sink uses a double base unit.

Gino and Louise want to have *Supaline* units fitted. This diagram shows a plan of their kitchen.

On squared paper draw the kitchen using a scale of 1:20. Design a kitchen for Gino and Louise using the units above. You should leave space for a dining area, and include a washing machine, fridge and a cooker (500 mm). Finally, work out the total price of your design.

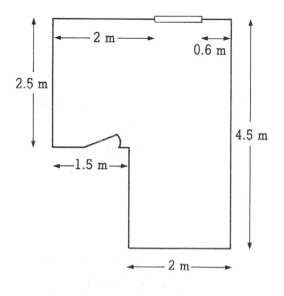

Angle facts 1

Information ⤵ Angles are used a lot in surveying. It is important that you know and can apply basic angle facts.

Here are 3 angle facts.

1   Angles which form a straight line add up to 180°.

2   Angles at a point add up to 360°.

3   The angles in a triangle add up to 180°.

Examples   **Using the 3 facts**

Calculate the missing angle in these diagrams.

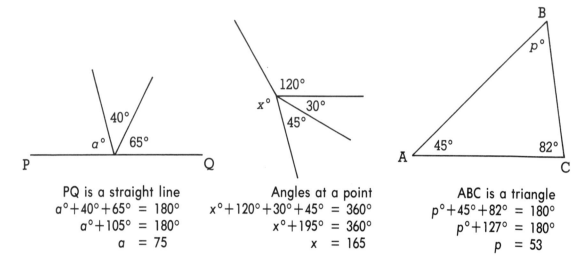

PQ is a straight line
$a° + 40° + 65° = 180°$
$a° + 105° = 180°$
$a = 75$

Angles at a point
$x° + 120° + 30° + 45° = 360°$
$x° + 195° = 360°$
$x = 165$

ABC is a triangle
$p° + 45° + 82° = 180°$
$p° + 127° = 180°$
$p = 53$

Exercise 6   Use the 3 facts to answer these questions. For each question you should

(a) sketch the diagram.

(b) write down the fact you are using.

(c) find the size of the missing angle(s).

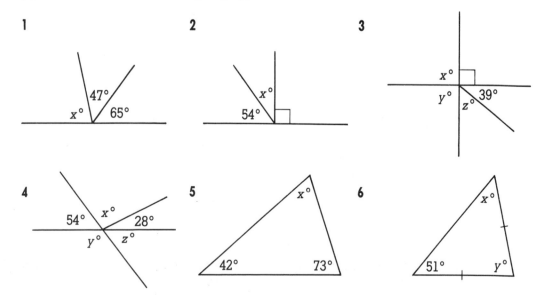

Angle facts 2

Information ↩   Here are 3 more angle facts which apply to lines which cross and lines which are parallel.

1   When 2 straight lines cross, the opposite angles are equal. These are called **vertically opposite angles.** These angles make an **X** shape.

2   AB and CD are parallel lines. Any straight line crossing the parallel lines is called a **transversal**. EF is a transversal.

The 2 shaded angles are equal. They are in corresponding positions in the diagram. They are called **corresponding angles.**

The letter **F** can be drawn round the angles. The F may have to be turned round to fit over the angles.

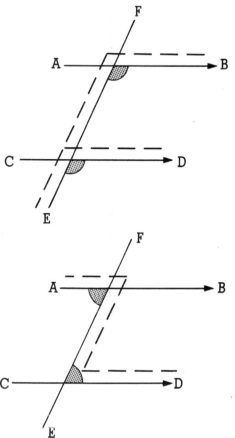

3   AB and CD are parallel lines. EF is a transversal.

The 2 shaded angles are equal. Their positions are on alternate sides of the transversal. They are called **alternate angles.**

The letter **Z** can be drawn round the angles. The Z may have to be turned round to fit over the angles.

Examples   **Using the facts**

Calculate the missing angles in these diagrams.

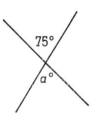

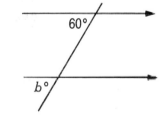

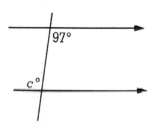

Vertically opposite (X)
$a = 75$

Corresponding (F)
$b = 60$

Alternate (Z)
$c = 97$

Exercise 7   Use the angle facts to answer these questions. For each question you should

(a) sketch the diagram.

(b) write down the fact you are using.

(c) calculate the size of the marked angles.

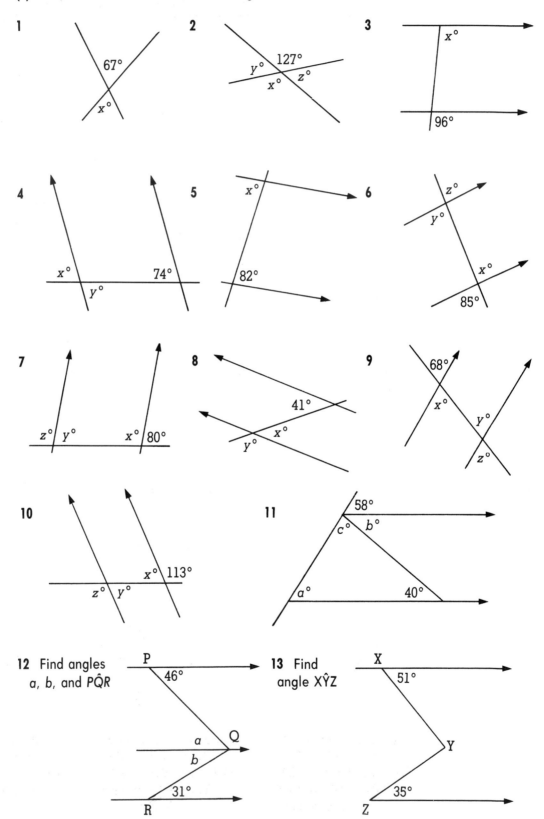

12 Find angles
a, b, and PQ̂R

13 Find
angle XŶZ

Angle facts 3

Information ⮌   The third set of angle facts are concerned with angles involved in circles. You should already be familiar with the words radius, diameter and circumference. The diagram opposite shows these.

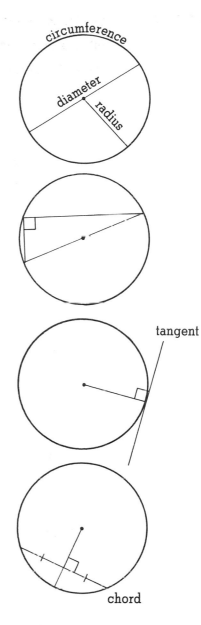

1   The angle in a semi-circle is a right-angle.

2   A radius meets a **tangent** at right-angles.

   A tangent is a straight line which touches the circumference of a circle at one point only, no matter how far the line is extended.

3   A **chord** is a straight line joining 2 points on the circumference of a circle. A radius meets a chord at right-angles if the chord is cut in half, or **bisected**.

Examples   Calculate the missing angles in these diagrams.

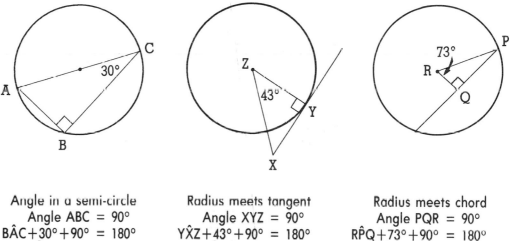

Angle in a semi-circle	Radius meets tangent	Radius meets chord
Angle ABC = 90°	Angle XYZ = 90°	Angle PQR = 90°
BÂC+30°+90° = 180°	YX̂Z+43°+90° = 180°	RP̂Q+73°+90° = 180°
BÂC+120° = 180°	YX̂Z+133° = 180°	RP̂Q+163° = 180°
BÂC = 60°	YX̂Z = 47°	RP̂Q = 17°

Exercise 8

Use these angle facts to answer the questions below. For each question you should

(a) sketch the diagram.

(b) write down which fact you are using.

(c) calculate the size of the missing angles.

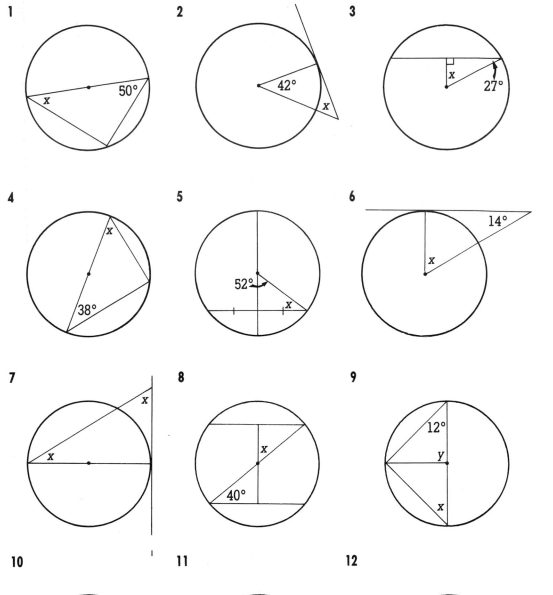

1

2

3

4

5

6

7

8

9

10

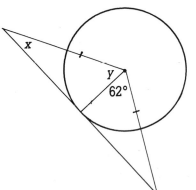

11

12

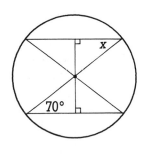

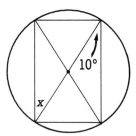

A tall story

The engineer in charge of the demolition used an instrument called a theodolite to accurately measure the angle of elevation to the top of the chimney. He was standing 100 metres from the foot of the chimney and found the angle to be 39°. The theodolite was positioned 1.6 m above the ground.

1 **Choose a suitable scale factor.**
Scale 1:2000
(1 cm represents 20 m)

2 **Change distance to scale size.**
100 m = 10 000 cm
10 000 cm ÷ 2000 = 5 cm

3 **Make an accurate scale drawing.**

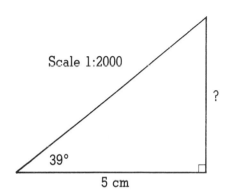

Scale 1:2000

?

39°

5 cm

4 **Measure the unknown side on the scale drawing.**
The vertical side of the right-angled triangle represents the height of the chimney. It is 4.0 cm long.

5 **Change from scale height to real height.**
$$4.1 \text{ cm} \times 2000 = 8200 \text{ cm}$$
$$8200 \text{ cm} = 82 \text{ m}$$
$$82 \text{ m} + 1.6 \text{ m} = 83.6 \text{ m}$$

The chimney is 83.6 metres high. The nearest building is 120 m away. This is more than 30 m from where the chimney is expected to fall so it is safe to use explosives to demolish it.

Investigations

1 The sketch shows a tower built on top of a mound. You cannot accurately measure the distance to the foot of the tower. How could you find its height?

2 You are standing on top of a cliff above a beach. In the distance is a lighthouse which you know is 1.3 km away. Explain how you could find the height of the cliff.

2 Work and pay

Do you remember? ⬅ In this unit it is assumed that you know how to:

1 multiply and divide by common mixed numbers using a calculator
2 work out a percentage of a sum of money
3 add and subtract time.

Exercise 0 will help you check. Can you answer all the questions?

Exercise 0 **Using a calculator to multiply and divide common mixed numbers**

1 $40 \times 1\frac{1}{2}$ 2 $24 \times 1\frac{1}{2}$ 3 $36 \times 1\frac{1}{4}$ 4 $68 \times 1\frac{1}{4}$

5 £1.20 $\times 1\frac{1}{2}$ 6 £2.56 $\times 1\frac{1}{2}$ 7 £2.72 $\times 1\frac{1}{4}$ 8 £3.48 $\times 1\frac{1}{4}$

9 $27 \div 1\frac{1}{2}$ 10 $62 \div 15\frac{1}{2}$ 11 $63 \div 5\frac{1}{4}$ 12 $175 \div 8\frac{3}{4}$

13 £34.10 $\div 15\frac{1}{2}$ 14 £52.54 $\div 35\frac{1}{2}$ 15 £83.85 $\div 32\frac{1}{4}$ 16 £66.40 $\div 20\frac{3}{4}$

Working out a percentage of a sum of money

17 Work out 5% of £20   18 Work out 8% of £45

19 Work out 15% of £380   20 Work out 32% of £750

21 Work out 12% of £36.50   22 Work out 15% of £328.40

23 Work out $6\frac{1}{2}$% of £560   24 Work out $7\frac{1}{4}$% of £48

Adding and subtracting time

Add up the following lists of times.

25 Hrs	Mins	26 Hrs	Mins	27 Hrs	Mins	28 Hrs	Mins
5	10	3	35	8	45	6	25
8	25	8	42	7	35	3	38
4	12	7	06	8	09	5	16
6	12	8	09	9	15	6	35

29 How long does a film last if it starts at 19:30 and ends at 21:39?

30 A train journey starts at 10:36 and ends at 15:45. How long is the journey?

31 A school day is from 08:45 until 15:40. How long are the pupils at school?

32 The pupils get a 15 minute break and a 55 minute lunch interval. How long do they spend in class?

Applying for jobs

Syeda and Mark have just left school and are looking for jobs. They buy the local paper every evening and discover that there are a lot of vacancies for sales people. They both enjoy meeting people and would like to try this. Here are 2 adverts which interest them.

SALES PERSON for busy clothes store 40 hour week. Salary £7000 p.a. plus 5% commission on annual sales over £6,000.
No previous sales experience necessary.
Apply in writing to
The Manager, Hurries, High Street.

YOUNG PERSON wanted as Sales Person in Clothes Dept of large store.
Must be energetic and hard working.
No previous sales experience necessary.
Salary £5500 p.a. plus
20% commission on all sales.
Apply in writing to
The Manager, Togs, White Hill

Syeda and Mark are trying to work out which job is likely to be the better paid. Which job would you apply for?

This unit will help you to solve the problem. It also explains about earning money, wage slips and take-home pay.

Pay slips

Information ⟿ In some jobs people are paid a **weekly wage** and in others they are paid a **monthly salary**. Each pay day they get a **pay slip**.

Example Graham leaves school and takes a job which pays £6000 per year, or £6000 **per annum**. He is paid monthly. As there are 12 months in a year, he expects to take home £500 each month. At the end of his first month he gets a surprise. Here is Graham's wage slip.

NAME G. Anderson		PERIOD TO 19/9/87	TAX CODE 242L	N.I. NUMBER YF536271D
BASIC £500	OVERTIME £	BONUS £	COMMISSION £	GROSS PAY £500.00
INCOME TAX £89.38	NAT. INS. £45.00	PENSION £	OTHER £	TOT DEDUCTIONS £134.38
				NET PAY £365.62

Graham gets £365.62 take-home pay, not the £500 he was expecting. He has a close look at his wage slip.

He spots that NET PAY = GROSS PAY − TOT DEDUCTIONS

$$£365.62 = £500.00 - £134.38$$

He also notices that TOT DEDUCTIONS = INCOME TAX + NAT. INS.

$$£134.38 = £89.38 + £45.00$$

Information ⟿ **Gross Pay** is your total earnings and includes basic pay plus any bonus, overtime or commission. Graham's gross pay is £500.

Total Deductions is the total amount to be taken away from your wages and includes Income Tax, National Insurance, pension and others. For Graham this is £134.38. This is what he forgot about.

Net pay is your take-home pay. Graham's net pay is £365.62.

We often say someone **earns** so much per week. We take this to mean their **gross pay**.

Example

Wage slips

From the wage slip calculate the missing amounts for

(**a**) gross pay   (**b**) total deductions   (**c**) net pay

NAME C. Tennant		PERIOD TO 25/4/87	TAX CODE 242L	N.I. NUMBER YW672398D
BASIC £130.00	OVERTIME £25.50	BONUS £	COMMISSION £	GROSS PAY £
INCOME TAX £35.08	NAT. INS. £13.99	PENSION £1.75	OTHER £	TOT DEDUCTIONS £
				NET PAY £

Gross pay: £130.00 + £25.50 = £155.50
Total deductions: £35.08 + £13.99 + £1.75 = £50.82
Net pay: £155.50 − £50.82 = £104.68

Exercise 1

For each wage slip find

(**a**) gross pay   (**b**) total deductions   (**c**) net pay

1

NAME B. Millar		PERIOD TO 15/7/86	TAX CODE 200L	N.I. NUMBER YW862982T
BASIC £98.00	OVERTIME £34.56	BONUS £	COMMISSION £	GROSS PAY £
INCOME TAX £28.20	NAT. INS. £11.93	PENSION £	OTHER £1.45	TOT DEDUCTIONS £
				NET PAY £

2

NAME A. Quicksand		PERIOD TO 17/9/87	TAX CODE 422H	N.I. NUMBER RT236590E
BASIC £625.00	OVERTIME £	BONUS £50.00	COMMISSION £164.00	GROSS PAY £
INCOME TAX £146.08	NAT. INS. £75.51	PENSION £	OTHER £	TOT DEDUCTIONS £
				NET PAY £

3

NAME M. Margiotta		PERIOD TO 23/12/87	TAX CODE 379H	N.I. NUMBER YW 512876D
BASIC £176.00	OVERTIME £66.00	BONUS £	COMMISSION £	GROSS PAY £
INCOME TAX £50.71	NAT. INS. £21.78	PENSION £5.50	OTHER £1.45	TOT DEDUCTIONS £
				NET PAY £

Basic pay

Information ⮎   In many jobs you work a standard number of hours, called a **basic week**, and you get paid a fixed **rate per hour**. The amount you earn in a week is called your **basic pay**.

Basic pay   = Rate per hour × Hours worked
Rate per hour = Basic pay ÷ Hours worked

Example   **Calculating basic pay**

Pete works on a production line. He gets paid £2.14 an hour for a 38 hour week. What is his basic pay?

Basic pay is £2.14 × 38 = £81.32

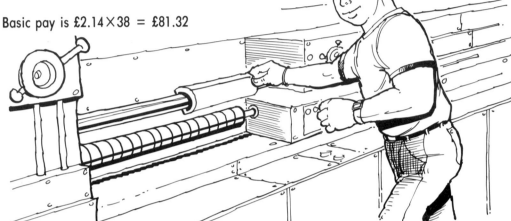

Example   **Calculating the rate per hour**

Omar gets £81.00 for working 37½ hours. How much is this per hour?

$$£81.00 ÷ 37.5 = £2.16$$

Exercise 2   Copy and complete the table below.

	Rate per hour	Hours	Basic pay
	£1.25	30	£1.25 × 30 = £37.50
1	£1.75	35	
2	£2.67	40	
3	£2.56	37	
4	£2.89	32	
5	£2.00	38	
6	£2.50	32½	
7	£1.80	35½	
8	£1.68	37¼	
9	£2.24	20¼	
10	£1.72	33¾	
11		30	£59.40
12		32	£72.00
13		34	£83.30
14		37	£119.14
15		37½	£106.50
16		26¼	£50.40

Overtime

Information ⇨ When you work more than your basic week you are paid **overtime**. This is usually paid at an increased rate per hour. The most common rate is **time-and-a-half**. This means that for every hour worked you are paid at 1½ times your basic rate. Other common rates are **time-and-a-quarter, time-and-a-third and double time**.

Example **Calculating overtime rate and pay**

Linda is a car mechanic. The garage is very busy and she often works overtime. Her basic rate is £3.14 per hour for a 38 hour week.

(a) Calculate her basic pay.

(b) Overtime is paid at time-and-a-half. Calculate her overtime rate.

(c) What is her gross pay if she works 5 hours overtime?

(a) Basic pay: £3.14 × 38 = £119.32

(b) Overtime rate: £3.14 × 1½ = £4.71 per hour

(c) Gross pay = Basic pay + Overtime pay
Overtime pay: £4.71 × 5 = £23.55
Gross pay: £119.32 + £23.55 = £142.87

Exercise 3 Calculate the 4 different overtime rates for each of these basic rates of pay.

	Basic rate	Time-and-a quarter	Time-and-a third	Time-and-a half	Double time
1	£1.20	£1.50			
2	£1.80			£2.70	
3	£2.04				£4.08
4	£2.28		£3.04		
5	£2.52				
6	£2.64				
7	£3.00				
8	£3.12				

9 Clair is a shop assistant and earns £1.80 per hour for a 40 hour week. She is paid overtime at time-and-a-quarter. Calculate her gross pay for a week in which she works 46 hours.

10 Bert, a plumber, is paid £3.12 per hour and time-and-a-half for hours worked over his basic 32 hour week. Calculate his gross pay for a week in which he works 40 hours ?

Time sheets

Information ⇨ Time sheets are sometimes used to keep a record of when employees start work, have lunch and stop for the day. Employees clock in and out and their times are recorded to the nearest 5 minutes.

Example

Calculating gross pay from a time sheet
You found in the previous example that Linda's basic pay is £119.32 for a 38 hour week and that her overtime rate is £4.71 per hour. Here is Linda's time sheet for another week.

(a)   How many hours did Linda work?

(b)   How much overtime did she work?

(c)   What is her gross pay?

Name L. Davies			Date 17/9/87		
	IN	OUT	IN	OUT	HOURS
MON	08:00	12:00	12:45	16:30	7:45
TUE	08:00	12:00	12:45	16:30	7:45
WED	08:00	12:00	12:45	17:30	8:45
THU	08:00	12:00	12:45	16:30	7:45
FRI	08:00	12:00	12:45	18:00	9:15
SAT	08:00	12:00			4:00
SUN					0:00

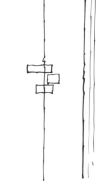

(a)   Remember that 1 hour = 60 minutes.

Add the minutes first
45+45+45+45+15 = 195 minutes
195 minutes = 3 hours 15 minutes

Hours worked Mon 7:45
Tue 7:45
Wed 8:45
Thu 7:45
Fri 9:15
Sat 4:00
³
45:15

(b)   Overtime: 45:15−38:00 = 7:15
7 hrs 15 mins = 7.25 hrs

(c)   Gross pay = Basic pay+Overtime
Basic pay: £3.14×38 = £119.32
Overtime: £4.71×7.25 =   £34.15
Gross pay: £119.32+£34.15 = £153.47

Exercise 4   For each of the following complete the table and find: (**a**) the total hours worked (**b**) the hours of overtime, (**c**) the overtime pay and (**d**) the gross pay. Set out your working as shown in the example on p.25.

1   Basic week: 40 hours
    Basic rate: £1.80 per hour
    Overtime rate: time-and-a-half

Name S. Ashton		Date 23/5/88		
	IN	OUT	IN	OUT	HOURS
MON	08:00	12:30	13:00	16:30	8:00
TUES	08:00	12:30	13:00	17:00	8:30
WED	08:00	12:30	13:00	18:30	
THU	08:00	12:30	13:00	18:30	
FRI	08:00	12:30	13:00	17:00	
SAT					
SUN					

2   Basic week: 40 hours
    Basic rate: £2.20 per hour
    Overtime rate: time-and-a-half

Name A. Daykin		Date 16/3/88		
	IN	OUT	IN	OUT	HOURS
MON	08:30	12:30	13:00	17:00	8:00
TUE	08:30	12:30	13:00	17:30	
WED	08:30	12:30	13:00	18:00	
THU	08:00	12:30	13:00	17:00	
FRI	08:30	12:30	13:00	17:00	
SAT	09:00	12:00			
SUN					

3   Basic week: 37½ hours
    Basic rate: £1.92 per hour
    Overtime Mon to Sat: time-and-a-third
    Overtime Sunday: double time

Name C. Archer		Date 03/2/88		
	IN	OUT	IN	OUT	HOURS
MON	8:30	12:00	12:30	4:30	7:30
TUE	8:30	12:00	12:30	4:30	7:30
WED	8:30	12:00	12:30	5:30	
THU	8:00	12:00	12:30	4:30	
FRI	8:00	12:00	12:30	5:00	
SAT	9:00	12:30			
SUN	9:00	12:00			

4   Basic 40 hour week
    Basic rate: £1.96 per hour
    Overtime Mon to Sat: time-and-a-half
    Overtime Sunday: double time

Name T. Wallington		Date 26/8/88		
	IN	OUT	IN	OUT	HOURS
MON	8:00	12:00	12:45	4:45	
TUE	8:30	12:00	12:45	5:30	
WED	8:00	12:00	12:45	4:45	
THU	8:00	12:00	12:45	6:00	
FRI	8:00	12:00	12:45	5:00	
SAT					
SUN	8:30	12:30			

Commission

Information ⟾ **Commission** is usually paid to people who sell things. The more they sell the more they will be paid. Their commission is normally a percentage of the value of their total sales.

Example
Calculating commission with no minimum sales
Sally is a check-out operator who is paid £100 per week plus 4% commission on all sales. Find her commission and gross pay if she sells £1250 worth of goods.

Commission: 4% of £1250 = £50
Gross pay = Basic pay+Commission
Gross pay: £100+£50 = £150

Example
Calculating commission with minimum sales
Andrea Payne, a double-glazing salesperson, gets £60 per week plus 12% commission on sales over £2000. Find her commission and gross pay for a week in which she sold £3500 worth of glazing.

Commission: 12% of (£3500−£2000)
            12% of £1500 = £180
Gross pay: £60 + £180 = £240

Exercise 5
Calculate the commission and gross pay earned by the following people.

1   H. Dancer earns £60 per week plus 12% of all sales. His sales this week are £1400.

2   R. Chaula earns £55 per week and 15% on sales over £1500. Her sales this week are £2450.

3   D. Hussain earns £45 plus 8% on all sales. Her sales this week amount to £2100.

4   Mrs Kay works for a mail order company. She receives no basic wage but gets 10% commission on all sales. How much will she earn if her sales are £560 ?

5   C. Singh sells shoes. He earns £67.87 per week plus 8½% commission on sales over £600. In one week he sells shoes to the value of £1180.

6   A. Payne sells double and triple glazing. She gets £68 per week plus 1% commision on all double glazing sales and 1½% on all triple glazing sales. One week her sales figures are: double glazing £5800, triple glazing £3300. Calculate her gross pay.

Estimating net pay

Information ↩ You have already seen that take home pay is less than gross pay. As a rough guide you can expect to have about one third of your gross pay deducted.

Example **Estimation of net pay**
Nadeen is a trainee chef. How much can she expect to take home if she is paid £2.35 per hour for a 40 hour week?

Total gross pay: £2.35 × 40 = £94
Approx. deductions
(to the nearest £1):   ⅓ of £94 = £31
Approx. net pay: £94 − £31 = £63

Note: Because you are estimating, answers are given to the nearest £1.

Exercise 6 Calculate the approximate net pay for each of the following.

1   Richard is a chef earning £87 per week.

2   Wendy is a trainee journalist earning £78 per week.

3   Omar is a receptionist earning £68.40 per week.

4   Elias is a printer earning £546 per month.

5   Pamela is a van driver earning £264 per month.

6   David is a painter earning £3.48 per hour for a 40 hour week.

7   Susan is a gardener earning £2.67 per hour for a 35 hour week.

8   Shabana is a clerk earning £3.06 per hour for a 37½ hour week.

Look at each of these job adverts and work out the approximate pay which you would expect to take home after deductions.

9

OPTICAL RECEPTIONIST

required at our new practice in the Asda Shopping Centre. Flexible working hours a must. 38-hour working week. Reception experience preferred. Age 18-25 years. Salary £6500 per annum.

Phone for appointment
669 6121

10

COOK
REQUIRED

To cook Chinese and European dishes. £85.00, 39-hour week. Food and accommodation provided.
EXPERIENCE REQUIRED

Tel: Bristol 82603

11

Assistant Play Leader
FOR SCOTTISH ADVENTURE PLAYGROUND ASSOCIATION FOR HANDICAPPED CHILDREN

Linn Park Adventure Playground Glasgow requires Assistant Play Leader with experience of Play and or Working with handicapped children. Practical skill and resourcefulness are necessary. Salary £5,500 pa increasing by annual increments.

Apply in writing enclosing brief CV to.
**SIMON PETERSON
51 WESTBOURNE GARDENS
GLASGOW G16**
by Saturday December 13

Investigation Look at some job adverts in your local paper. Choose the jobs in which you are interested and estimate the weekly or monthly net pay in each case.

National insurance

Information ⇨ National Insurance (N.I.) is paid by both you (the **employee**) and your **employer**. It helps to pay for :

National Health Service   Pensions
Statutory sick pay   Unemployment benefits

The amount paid in National Insurance depends on your gross wage. You pay 9% of your gross pay and your employer pays 10.45%. These percentages can be calculated but to simplify things the DHSS provide tables which give the contributions for both the employee and employer.

Example   **Calculating your N.I. contribution**

Look again at Graham's payslip.

NAME G. Anderson		PERIOD TO 19/9/87	TAX CODE 242L	N.I. NUMBER YF536271D
BASIC £500	OVERTIME £	BONUS £	COMMISSION £	GROSS PAY £500.00
INCOME TAX £89.38	NAT. INS. £45.00	PENSION £	OTHER £	TOT DEDUCTIONS £134.38
				NET PAY £365.62

Here is part of the table which affects Graham.

Gross pay	Total of employee's and employer's contributions payable	Employee's contribution payable	Employer's contribution*
£	£	£	£
497.00	96.86	44.82	52.04
499.00	97.25	45.00	52.25
501.00	97.64	45.18	52.46
503.00	98.03	45.36	52.67

£500 is not shown in the first column so you take the next smallest figure, £499.
Total contribution: £97.25
Employee's contribution: £45
Employer's contribution: £52.25

Exercise 7

Use the National Insurance tables below to find

(a) the employee's contribution

(b) the employer's contribution

(c) the total contribution for the following wages.

If the exact gross pay figure is not shown in the table, use the next smaller figure shown.

Gross pay	Total of employee's and employer's contributions payable	Employee's contribution payable	Employer's contribution*	Gross pay	Total of employee's and employer's contributions payable	Employee's contribution payable	Employer's contribution*	Gross pay	Total of employee's and employer's contributions payable	Employee's contribution payable	Employer's contribution*
£	£	£	£	£	£	£	£	£	£	£	£
83·50	16·29	7·54	8·75	93·50	18·24	8·44	9·80	103·50	20·18	9·34	10·84
84·00	16·38	7·58	8·80	94·00	18·33	8·48	9·85	104·00	20·27	9·38	10·89
84·50	16·49	7·63	8·86	94·50	18·43	8·53	9·90	104·50	20·38	9·43	10·95
85·00	16·58	7·67	8·91	95·00	18·52	8·57	9·95	105·00	20·47	9·47	11·00
85·50	16·68	7·72	8·96	95·50	18·63	8·62	10·01	105·50	20·57	9·52	11·05
86·00	16·77	7·76	9·01	96·00	18·72	8·66	10·06	106·00	20·66	9·56	11·10
86·50	16·88	7·81	9·07	96·50	18·82	8·71	10·11	106·50	20·77	9·61	11·16
87·00	16·97	7·85	9·12	97·00	18·91	8·75	10·16	107·00	20·86	9·65	11·21
87·50	17·07	7·90	9·17	97·50	19·01	8·80	10·21	107·50	20·96	9·70	11·26
88·00	17·16	7·94	9·22	98·00	19·11	8·84	10·27	108·00	21·05	9·74	11·31
88·50	17·26	7·99	9·27	98·50	19·21	8·89	10·32	108·50	21·15	9·79	11·36
89·00	17·36	8·03	9·33	99·00	19·30	8·93	10·37	109·00	21·25	9·83	11·42
89·50	17·46	8·08	9·38	99·50	19·40	8·98	10·42	109·50	21·35	9·88	11·47
90·00	17·55	8·12	9·43	100·00	19·50	9·02	10·48	110·00	21·44	9·92	11·52
90·50	17·65	8·17	9·48	100·50	19·60	9·07	10·53	110·50	21·54	9·97	11·57
91·00	17·75	8·21	9·54	101·00	19·69	9·11	10·58	111·00	21·64	10·01	11·63
91·50	17·85	8·26	9·59	101·50	19·79	9·16	10·63	111·50	21·74	10·06	11·68
92·00	17·94	8·30	9·64	102·00	19·89	9·20	10·69	112·00	21·83	10·10	11·73
92·50	18·04	8·35	9·69	102·50	19·99	9·25	10·74	112·50	21·93	10·15	11·78
93·00	18·13	8·39	9·74	103·00	20·08	9·29	10·79	113·00	22·02	10·19	11·83

1 Barry earns £98 per week.

2 Raschid earns £113 per week.

3 Jill earns £100.50 per week.

4 Sureta earns £89.40 per week.

5 Avril earns £112.75 per week.

6 Jane earns £2.55 per hour for a 40 hour week.

7 Sajid earns £2.85 for a 35 hour week.

8 Karen earns £3.54 for a 28 hour week.

9 Scott earns £4.28 per hour for a 22½ hour week.

10 Margaret earns £2.84 per hour for a 37½ hour week.

11 Darren earns £140 per week. How much will he pay in N.I. contributions?
Note: this cannot be found from the tables but we know that the employee's contribution is 9% of his gross pay.

Allowances

Information ⤳   **Income tax** is paid to the Government. Everyone is allowed to earn a certain amount of money each year on which they do not pay income tax. This is called their **allowance**. Allowances depend on individual circumstances, such as whether you are single or married, if you look after a dependant relative, and other personal circumstances.

Details of allowances can be found in leaflet IR22 available from tax offices. In 1987 the single person's allowance was £2425 and the dependant relative allowance was £100.

Taxable income is the amount of money on which income tax is paid.

Taxable income = Annual gross pay — Total allowances

Examples   **Calculating an allowance**
Duncan is single and looks after his elderly mother. What are his allowances?

Single person:	£2425
Dependant relative:	£ 100
Total allowance:	£2525

Calculating taxable income
Find Graham's taxable income if he earns £500.00 per month and his total allowances are £2425 per year.

Annual gross pay: £500×12 = £6000
Taxable income = Annual gross pay — Total allowances
£6000−£2425 = £3575

Exercise 8   Calculate the total allowance and taxable income for the following people.

1   ·Mr Black is single. He earns £9000 per year.

2   Ms Peters is a single woman who looks after her elderly mother. She earns £11 500 per year.

3   Mike is single, earns £14 000 and looks after his ageing father.

4   Tricia is single and lives alone. She earns £12 475 per year.

5   Copy and complete this table.

	Wage or salary	Annual gross pay	Total allowances	Taxable income
	£ 95 per week	£95×52 = £4940	£2425	£4940−£2425 = £2515
6	£103.50 per week		£3155	
7	£143.35 per week		£3765	
8	£546 per month		£4365	
9	£624 per month		£3243	
10	£485 per month		£2265	
11	£104.65 per week		£3342	
12	£647 per month		£4673	

Information ↪ If a married couple both work, the husband receives the married man's allowance, but the wife receives only the single person's allowance. The married man's allowance in 1987 was £3795.

Example **Calculating an allowance**
Dave and Jameda are married and have 2 young children. Jameda works part-time. Dave also gets an allowance of £265 for a home improvement loan. What is his total allowance?

Married man's allowance:	£3795
Improvement loan allowance:	£ 265
Total allowance:	£4060

Note: No tax allowance is presently given for children.

Exercise 9 Calculate the total allowances and the taxable income of the following people.

1 Mr Granger, who is married with 1 child. He earns £8500 per year

2 Mr and Mrs Davies have 3 children. They both work and Mr Davies has additional tax allowances totalling £862. He earns £9215 and Mrs Davies earns £10 657.

3 Mr Duncan is married and earns £12 319. Mrs Duncan does not work. Mr Duncan has a home improvement loan for which he gets a tax allowance of £537.

4 Mr and Mrs Chawla have 1 child. Mr Chawla earns £13 422 and has additional tax allowances totalling £428. Mrs Chawla earns £9600.

5 Mr and Mrs Edwards run their own business. They each earn a salary of £12 000, and have no additional tax allowances.

Information ↪ When you start work you will be given a **tax code**. This will appear on your payslip, and tells you about your tax allowance. Tax codes are in 2 parts — a number followed by a letter.

The number is simply the first 3 digits of the tax allowance.

The letter tells you the main allowance. **L** shows that the code includes a single person's allowance. **H** shows the code includes a married man's allowance.

Example **Tax codes**
As we have seen Dave has a tax allowance of £4060 and is married. What is his tax code?

Tax code is the first 3 figures of his allowance followed by the letter H as he is married.

His tax code is 406H

Exercise 10 1 – 5 Write down the tax code for each of the people in the last exercise.

Here are 4 employees' tax codes. What do they tell you?

6 Sue 242L   7 Joe 342H   8 Shabana 245L   9 Omar 402H

Income tax

Information ⤸ The amount you pay in income tax is calculated as a percentage of your taxable income. This amount sometimes changes after a budget. In this unit you should assume that most people pay 30% of their taxable income. This is called the **basic rate of income tax**. You should also check up the current rate of income tax. In some job adverts you may see **per annum**. Remember this simply means for a year.

Example **Calculating income tax**
How much does Graham pay in income tax if he earns £6000 per year and has total allowances of £2425 ?

Taxable income: £6000−£2425 = £3575
Income tax: 30% of £3575 = £1072.50

Exercise 11 Calculate (**a**) the taxable income
(**b**) the income tax to be paid for each of the following.

1 David earns £4940 per year. His total allowances are £2425.

2 Susan earns £5450 per year. Her total allowances are £2654.

3 Stewart earns £6025 per year. His total allowances are £3155.

4 Julie earns £5832 per year. Her total allowances are £2775.

5 Waseem earns £6996 per year. His total allowances are £3657.

6 Shabana earns £5148 per year. Her total allowances are £2746.

In the following problems calculate for 1 year
(**a**) the total allowances
(**b**) the taxable income
(**c**) the income tax to be paid.

7 Grant is a married man with 2 children. He earns £7000 per year, and his wife doesn't work.

8 Simon is a single man who earns £6888 per year. He looks after his elderly mother.

9 Robert is a married man who earns £7020 per year. He has a home improvement loan for which he gets an allowance of £385.

10 Jane is a single woman who earns £10 716 per year as an engineer. She pays an annual membership to the Institute of Mechanical Engineers for which she gets an allowance of £45. In addition she gets an allowance of £657 for a loan which she is repaying.

11 Paul is single and earns £8800 per year. He has a tax allowance of £904 for his home improvement loan, and other allowances totalling £125.

Information ⟿   Anyone whose taxable income is greater than £17 900 pays a **higher rate of income tax** on the excess income. Here are the first 3 rates.

Taxable income	Rate of tax
£1 to £17 900	30% (approx.)
£17 900 to £20 400	40%
Above £20 400	45%

Example

Calculating income tax for higher rates
Ms Lewis is a company director who earns £22 000 per year. Her total allowances are £3685. How much income tax must she pay?

First we need to work out her annual taxable income.
Annual taxable income = Annual gross pay − Total allowances
         £22 000−£3685 = £18 315

This taxable income takes her into the first band of the higher rate of tax (40%).

Income tax = 30% of £17 900 + 40% of (£18 315−£17 900)
         = 30% of £17 900 + 40% of £415
         = £5370 + £166
         = £5536

Exercise 12   Calculate the income tax to be paid for the following.

1   Mr Bashey earns £19 500 per year. His total allowances are £3867.

2   Ms Rogers earns £23 000 per year. Her total allowances are £4997.

3   Mr Turner earns £25 680 per year. His total allowances are £4386.

4   Ms Forbes earns £26 000 per year. Her total allowances are £5287.

5   Mr Hill earns £27 400 per year. His total allowances are £5823.

Information ⇨   All the calculations you have done involving income tax have been for a year. Most people, however, are paid weekly or monthly.

Example   **Calculating monthly income tax**
Here is Graham's pay slip.

NAME G. Anderson		PERIOD TO 19/9/87	TAX CODE 242L	N.I. NUMBER YF536271D
BASIC £500	OVERTIME £	BONUS £	COMMISSION £	GROSS PAY £500.00
INCOME TAX £89.38	NAT. INS. £45.00	PENSION £	OTHER £	TOT DEDUCTIONS £134.38
				NET PAY £365.62

Graham earns £500 per month. His total allowances are £2425. How is his income tax deduction of £89.38 calculated?

Monthly allowance: £2425 ÷ 12 = £202.08
Monthly taxable pay: £500 − £202.08 = £297.92
Monthly income tax: 30% of £297.92 = £ 89.38

Example   **Calculating weekly income tax**
Miriam is a plumber who is paid £3.60 per hour for a 35 hour week.

She is single but cares for her elderly father. Her annual tax allowance is £2525. How much will she pay in income tax if she works a basic week?

Gross pay: £3.60 × 35 = £126.00
Weekly tax allowance: £2525 ÷ 52 = £ 48.56
Taxable pay: £126.00 − £48.56 = £ 77.44
Income tax: 30% of £77.44 = £ 23.23

Exercise 13   For each of these people, calculate

(a)   their total gross pay.

(b)   their National Insurance.

(c)   their income tax.

(d)   their complete wage slip.

1   Brian Millar earns £98 per week. His overtime is £34.56. Apart from N.I. and income tax he also has £1.45 union fees deducted each week. Brian is single.

2   Amina Hussain is a sales representative earning £74 per week. Her commission is £43.76 and she has a bonus of £35. She is single and lives at home.

3   Maurice Metro is a driving instructor who is paid £570 per month. His annual tax allowance is £3864. Last December he earned a bonus of £75. In addition to N.I. and income tax he pays £50 per month to his employers to repay a company loan.

4   Andy Quicksand is the manager of a Do-It-Yourself superstore. He is paid £625 per month. He is married with 2 children and has an annual tax allowance of £4225. Last month his store had record sales and Andy earned £164 commission and a £50 bonus.

5   Mario Margiotta is an electrician earning £4.40 per hour for a 40 hour week. Overtime is paid at time-and-a-half. In one week he works 10 hours overtime. In addition to N.I. and income tax he also pays £1.45 per week union fees and £5.50 per week for a pension scheme. He is married with 2 children, whom his wife looks after.

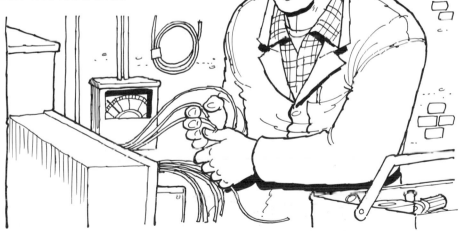

How do your wage slips for questions 1, 4 and 5 compare with the three in Exercise 1?

Investigation   1   Look at the job adverts from your local paper. Choose one job that you are interested in. Write out a pay slip for yourself, calculating your N.I. and income tax and net pay.

2   Write a letter applying for the job. What are the things you should say in the letter?

Applying for jobs

Look back at the 2 jobs Syeda and Mark are interested in. The only way to compare the 2 salaries is to find the gross pay for different annual sales.

Hurries £7000 plus 5% of sales over £6000 per year.
Togs   £5500 plus 20% of all sales.

Consider the salaries for sales starting at £1000 and going up in steps of £1000.

Annual sales of £1000
Hurries £7000 only as sales are not over £6000
Togs   £5500 + 20% of £1000
          = £5500 + 200
          = £5700

The table below gives the salaries for different values of sales. Copy and complete the graph.

Sales	Hurries	Togs
£1000	£7000	£5700
£2000	£7000	£5900
£3000	£7000	£6100
£4000	£7000	£6300
£5000	£7000	£6500
£6000	£7000	£6700
£7000	£7050	£6900
£8000	£7100	£7100
£9000	£7150	£7300

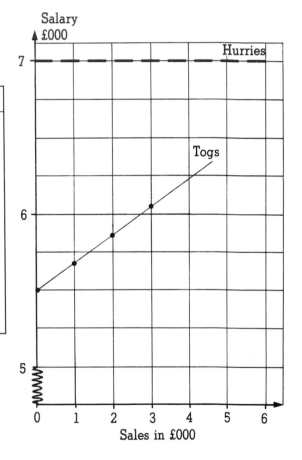

What value of sales gives the same pay for each method? What value of *monthly* sales is this?

Do you think that you would be able to achieve this value of monthly sales?

Do you think the value of sales will be constant each month?
Explain your answer. When do you think you will earn most?

Explain which job you would take and why?

What other factors besides your earnings would you consider?

Exercise 14   The questions below can all be solved in the same way as the job problem.

1   Ms. Barker is offered a choice of pay. Either a fixed wage of £175 gross per week, or £45 per week plus 5% commission on all sales. Copy and complete the table.

Sales	Commission	Gross pay
£500	£25	£70
£1000	£50	£95
£1500		
etc.		

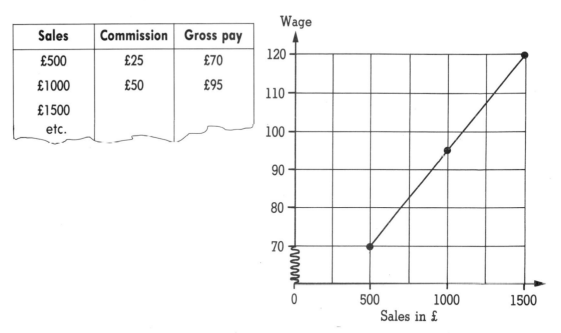

Copy and complete the graph above, plotting all the figures in your table. Estimate the value of sales Ms. Barker needs to make commission a better choice than the fixed wage?

2   Mr. Coates is offered a similar deal — £150 per week or £35 plus 6% commission on all sales. Copy and complete the table below.

Sales	Commission	Gross pay
£500	£30	£65
£1000	£60	
etc.		

Extend the table and draw a graph. How much does he have to sell before commission is the better deal?

3   After working for one year, Mr. Coates is offered a new deal — £200 per week or £50 plus 6% of commission on all sales. Complete a table and graph as in question 2. How much does he now have to sell before commission is a better deal?

4

> Salesperson wanted for large firm
> Good sales prospects
> £7500 + 3% of sales over £5000
> OR
> £5500 + 15% of all sales
> Apply in writing to
> Sales International, Grove St., Derby

Consider different values of sales and work out when each method is best. Draw a graph and estimate the value of sales which gives the same gross pay.

3 Symbolic maths

Do you remember? ⟿

In this unit it is assumed that you know how to:

1 order calculations
2 estimate the answer to a calculation
3 use your calculator correctly
4 use simple algebraic notation.

Exercise 0 will help you check. Can you answer all the questions?

Exercise 0

Order of calculations

Find the answer to these calculations.

1 $3 \times 6 - 4$ 2 $5 + 4 \times 2$ 3 $4 \times 7 - 3 \times 5$

4 $5 \times (7-3) + 6$ 5 $53 - 3 \times (6+3)$ 6 $36 \div 4 + 3 \times (19-5)$

Estimation and calculators

Estimate the answer to each of these calculations.

7 4.21×7.93 8 37.2×3.1431 9 $19.6 \div 3.725$

10 $322.4 \div 6.421$ 11 $19.43 \div 43.2$ 12 417×1.42

13 17.6×4.412 14 $\dfrac{56.21 \times 19.5}{28.71}$ 15 $\dfrac{36.7 \times 58.9}{27.2}$

16 $\dfrac{231 \times 9.61}{46.821}$ 17 $\dfrac{58.21 \times 3.12}{9.11}$ 18 $\dfrac{712 \times 36.2}{4423}$

19 $\dfrac{68.3 \times 1.63}{38.7}$ 20 $\dfrac{23.6 \times 4.73}{32.1 \times 4.11}$ 21 $\dfrac{5.27 \times 7.91}{2.43 \times 24.4}$

22 $\dfrac{744.21 \times 105.41 \times 12.74}{483.8 \times 74.32}$ 23 $\dfrac{76.4 \times 5.33 \times 295.4 \times 3.3}{562.5 \times 93.4 \times 13.7}$

24 Use your calculator to find more accurate answers to questions 7–23.

Notation

Simplify each of the following.

25 $a+a+a$ 26 $y+y+y+y+y$ 27 $8 \times p$ 28 $1 \times v$

29 $A \times C$ 30 $R \times T$ 31 $c \times d$ 32 $d \times d$

33 $y \times y$ 34 $s \times s$ 35 $t+t$ 36 $5 \times n \times n$

37 $2 \times 5y$ 38 $5 \times 3s$ 39 $8 \times 6z$ 40 $3 \times 4c^2$

41 Arrange the digits 1, 2, 3, 4 and 5 into two numbers so that their product is as large as possible. For example $123 \times 45 = 5535$.

Camp stool puzzle

One summer's evening during their camping holiday, Michelle and Cuong began to study the stars. They imagined joining the stars up and drawing simple objects in the sky, like a tent and camp stool. The diagram shows some of the stars that they saw. Here they are joined up to show how they are grouped in **constellations.**

Northern Hemisphere

Back at school, one wet September afternoon, Cuong thought back to their holiday. Then he began to draw some camp stools on the dotted paper in front of him.

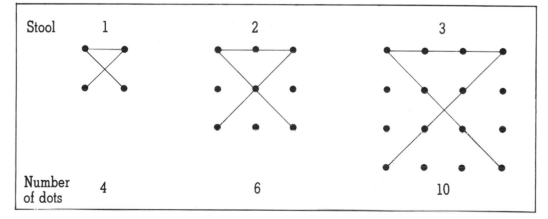

He continued with this for some time to see if he could see a pattern connecting the stool number with the number of dots it contained.

When he showed the puzzle to Michelle she said that stool 4 has 12 dots. Was she right? She could also tell him the number of dots in the 15th and 24th stools. How did she work this out?

This unit will show you how to solve problems like this without drawing the stools.

Like terms

Information ➪   If you have a lot to write you might try to think of a shorthand way of doing this. Algebra is a kind of mathematical shorthand where letters are used in a **formula**. Sometimes the same letter occurs twice or more in the same formula. The parts of the formula with the same letter are called **like terms**. In algebra you can often simplify formulae or expressions by identifying and collecting together like terms.

Examples   **Forming expressions and collecting like terms**
Lucy is a gardener who
works for the local council.
She is in charge of 2 parks
and has to order the flowers
for each park.

Here is her order

Park 1	25 roses	15 daffodils	18 geraniums
Park 2	12 roses	23 daffodils	10 geraniums
Total order	37 roses	38 daffodils	28 geraniums

When she was in the parks deciding the number of flowers to order Lucy used shorthand. Using *r* for roses, *d* for daffodils and *g* for geraniums her working looked like this.

Park 1	Park 2	Total
$25r + 15d + 18g$	$12r + 23d + 10g$	$37r + 38d + 28g$

Can you see how Lucy has calculated the totals?

$25r$ and $12r$ are **like terms.** They can be added together.
$25r$ and $15d$ are **unlike terms.** They cannot be added together.
She cannot add all the terms together and simply order a total of 103 flowers because she wants 3 different types and must keep the 3 types separate.

Examples   **Simplifying expressions by collecting like terms**

1   $2c + 5c = 7c$ 2 $3d + 7d - 4d = 6d$

3   $5p^2 + 9p^2 - 3p^2 = 11p^2$ 4 $2d + 3e + 5d + e = 7d + 4e$

5   $7x + 3y - 2x + 3y = 5x + 6y$ 6 $5d^2 + 3d - d^2 + 4d = 4d^2 + 7d$

7   $6st + 3st - 2st = 7st$ 8 $2pq + 3st - pq + 5st = pq + 8st$

Exercise 1

1 George has a large garden with 2 areas for growing vegetables. He decides to plant the following vegetables in each area.

Area 1 40 potatoes, 50 carrots and 20 leeks   or $40p+50c+20l$
Area 2 20 potatoes, 20 carrots and 45 leeks   or $20p+20c+45l$

Copy and complete the shorthand to find the total of each type of vegetable George will need.

2 Pauline is a baker and is ordering cakes for her two shops. Copy her shorthand and add together the like terms to calculate the total order.
Shop 1 40 doughnuts, 24 tarts and 30 cakes   or $40d+24t+30c$
Shop 2 36 doughnuts, 12 tarts and 24 cakes   or $36d+12t+24c$

3 Fiaz is responsible for stocking a hotel with drinks. He checks the number of bottles of fruit juice he has in stock in the 3 store rooms.

Room 1 16 pineapple,   5 orange   and 9 grapefruit
Room 2   5 pineapple, 12 orange   and 6 grapefruit
Room 3 18 pineapple,   8 orange   and 3 grapefruit

Use a suitable shorthand to write this information and calculate the totals.

4 A shopkeeper is checking the notes in her tills. Her rough working is

Till 1 12f and 4t, **Till 2** 11f and 5t, **Till 3** 14f and 3t

where f stands for a five pound note and t stands for a ten pound note. Write down a simple expression for her notes in all 3 tills. Add the like terms to find her total number of each type of note.

5 Marjory is a librarian. She is checking the number of books and magazines returned during the day. They are stored on 4 shelves.

Shelf 1 18 hardbacks, 19 paperbacks and 4 magazines.
Shelf 2   8 hardbacks, 16 paperbacks and 1 magazine
Shelf 3 23 hardbacks,   9 paperbacks and 0 magazines
Shelf 4   6 hardbacks, 12 paperbacks and 7 magazines.

Use a suitable shorthand and rewrite Marjory's 4 lists. Add your 4 expressions together to make one new list, and collect the terms. What is Marjory's total for each of the 3 types?

Simplify by adding or subtracting like terms.

6 $4a+5a$

7 $10x-3x$

8 $5k+3k-2k$

9 $6p-4p+2p-4p$

10 $2q+8q+10q$

11 $5r+7r+3s+6s$

12 $9x-4x+3y-2y$

13 $12c+5d-7c+6d$

14 $8f+2g+2f+6g$

15 $3a+6b+2b-2a+7c$

16 $12z+6y+2z+4y-3y$

17 $8h+4j-2h+3k-2j+4k$

18 $6v+8w-2t+3w-4v+6t$

19 $16s+21t-4s-3s-14t$

20 $14r+3t+6s+7r-2t-3s$

21 $4ab+3ab$

22 $3xy+2xy$

23 $6xy+2xy-3xy$

24 $4ab+3pq-2ab+7pq$

25 $7x^2+3y^2+9x^2+y^2$

26 $7v+5v^2+2v+3v^2$

27 $8x+3x^2+5x+2x^2$

28 $5c+2d+c^2-d+5c^2$

29 $5x^2+3y+2y^2+3y+2x$

Using brackets

Information ⮌ You are already familiar with the use of brackets in calculations. In algebra they are used to group terms together. Sometimes it is easier to do the calculation if the brackets are removed first. To remove brackets each term inside the bracket must be multiplied by the term outside.

Example **The use of brackets**
Lucy decides that her two gardens should have the same flower display. Each garden is to have 10 roses, 15 daffodils and 12 geraniums.

This time her shorthand is

$$2(10r+15d+12g) = 20r+30d+24g$$

Lucy has used brackets to group together the number of flowers for each garden, and has multiplied them all by 2.

$$2\times10r = 20r \qquad 2\times15d = 30d \qquad 2\times12g = 24g$$

Examples **Removing brackets**

1	$2(3x+5) = 6x+10$	2	$6(5a+4b) = 30a+24b$	3	$3(p+2q) = 3p+6q$
4	$4(5s-3t) = 20s-12t$	5	$7(5v-9) = 35v-63$	6	$4(3-2r) = 12-8r$
7	$3(2x+5y-z) = 6x+15y-3z$	8	$5(2a-3+4b) = 10a-15+20b$		

Exercise 2

1 George is a gardener. He wants to test 2 different types of fertiliser on 2 plots in his garden. He decides to plant the same number of vegetables in each of the 2 plots. He is going to plant 30 potatoes, 40 carrots and 35 leeks. His shorthand is

$$2(30p+40c+35l) =$$

Copy and complete his shorthand to find the total number of each type of vegetable George must buy.

2 Pauline stocks each of her 2 shops with 50 doughnuts, 20 tarts and 30 cakes. Her driver, Andy, is told to load the van with the correct number of each. Copy and complete Andy's shorthand to find the total number of doughnuts, tarts and cakes he will need.

$$2(50d+20t+30c) =$$

3 Mumtaz owns a shop which has 3 tills. Each morning he puts 10 pound coins, 4 five pound notes and 1 ten pound note in each till. Use brackets and a suitable shorthand to find the total number of each he will need.

4 Roberto owns 4 discos. In each he has 140 singles, 23 LPs and 6 tapes. Use brackets and shorthand to find the total number of singles, LPs and tapes he has.

5 Senga delivers newspapers to a chain of 5 shops. Each shop gets 30 *Guardians*, 40 *Records* and 35 *Expresses*. Use brackets and shorthand to find the number of each paper Senga has to collect each morning.

Remove the brackets from the following expressions by multiplying each term in the bracket by the number outside.

6 $2(x+3)$	**7** $3(y+5)$	**8** $6(2+z)$
9 $7(a+b)$	**10** $5(3c+2d)$	**11** $7(4f+2g)$
12 $9(5x+4)$	**13** $10(3y+2)$	**14** $2(3h+2j+4k)$
15 $8(4l+3p+2q)$	**16** $4(2a+3b+c)$	**17** $5(4+7m+4n)$
18 $6(2x-5)$	**19** $5(2c-7)$	**20** $8(3p-2q)$
21 $6(4g-5h)$	**22** $5(2c-3d-e)$	**23** $9(3p-6q-r)$
24 $4(3c-5d+e)$	**25** $7(5r+6s-2t)$	**26** $5(3u-2v-5w)$
27 $4(c-3d+2e-f)$	**28** $6(p+3q-5r-s)$	**29** $8(w-x-6y+2z)$

Information ⮑ Sometimes sets of brackets are linked by a + or − sign. You can simplify them by first multiplying out the brackets and then collecting like terms.

Example **Simplifying expressions with 2 or more brackets**

Simplify $3(2x+3y-z)+2(5x-y+5z)$

$$3(2x+3y-z)+2(5x-y+5z)$$

Multiply out the brackets: $6x+9y-3z+10x-2y+10z$

Group together like terms: $6x+10x+9y-2y+10z-3z$

Add or subtract like terms: $16x+7y+7z$

Note the change in the order of the z terms when the expression is rearranged. It is useful to put + terms before − terms.

Exercise 3 Simplify the following expressions. Set out your working as in the example above.

1 $2(2a+3b)+5(a+5b)$	**2** $3(8r+2s)+2(2r+5s)$	**3** $5(2x+5y)+4(x+y)$
4 $9(g+4h)+2(g+3h)$	**5** $4(5v+w)+2(v+2w)$	**6** $5(2s+t)+2(2s-t)$
7 $6(7x+3y)+2(3x-4y)$	**8** $7(3c+4d)+4(c-5d)$	**9** $5(5v+4w)+4(3v-4w)$
10 $4(c-2d)+5(c+3d)$	**11** $5(2x-y)+3(3x+4y)$	**12** $4(3p-4q)+7(p+3q)$

13 $3(2x+4y+2z)+5(x+2y+3z)$ **14** $2(a+3b+2c)+7(2a+3b+c)$

15 $4(p+2q+r)+5(2p+3q+2r)$ **16** $4(2r+s+t)+3(r+3s+2t)$

17 $5(3w+2x+y)+2(w+3x+y)$ **18** $2(3x+y+3z)+3(2x+2y-z)$

19 $4(a+3b+4c)+2(2a+b-3c)$ **20** $5(p+q+3r)+2(2p-q+2r)$

21 $3(c+2d+e)+2(3c-2d-e)$ **22** $4(2r+3s+2t)+2(5r-3s-3t)$

23 $7(2r+3s+t)+3(r-5s-2t)$ **24** $2(x+y-z)+3(2x+3y+2z)$

25 $5(5j+2k-l)+4(2j+k+2l)$ **26** $7(2r+3s-t)+3(r+2s+3t)$

27 $3(p-2q+r)+5(p+3q+2r)$ **28** $4(x-2y-z)+7(x+3y+z)$

29 $2(c-2d-3e)+5(c+d+3e)$ **30** $3(2r-2s-3t)+7(r+s+4t)$

31 $2(3r+2s)+4(r-s)+3(4r+2s)$ **32** $5(x-y)+3(2x+4y)+2(3y-2x)$

Information ↩ Brackets are also used to simplify expressions. Putting brackets into an expression is called **factorising**. The number which goes outside the bracket is called the **common factor**, and must divide into all the terms to be factorised.

Example

Factorising

Lucy has been given extra flowers to be evenly divided between her 2 parks. She has 50 roses, 30 daffodils and 18 geraniums. Here is her working to find how many of each flower is to be planted in each park.

$$50r + 30d + 18g = 2(25r + 15d + 9g)$$

This time Lucy has put in brackets and taken out a common factor of 2.

Examples

Factorising expressions

1 $4x + 12 = 4(x + 3)$ 2 $15 - 9x = 3(5 - 3x)$

3 $8a + 10b = 2(4a + 5b)$ 4 $6r - 10s + 4 = 2(3r - 5s + 2)$

5 $12pq - 18qr = 3q(4p - 6r) = 3q \times 2(2p - 3r) = 6q(2p - 3r)$

In the last example you found the **highest common factor** to put outside the bracket. This means that the part inside the bracket cannot be factorised again. When you have factorised an expression you should always check to see if it can be factorised again.

Exercise 4

1 George saw this special offer for ready-to-plant vegetables, in his local garden centre.

BARGAIN PACK
READY-TO-PLANT
~
90 Potatoes
70 Carrots
52 Leeks

George buys the pack and decides to split the vegetables equally between his 2 plots. Copy and complete the expression below to find the number of each vegetable George should plant in each plot.

$$90p + 70c + 52l = 2(\qquad\qquad)$$

2 One morning Pauline is short in her baking. She only has 90 doughnuts, 34 tarts and 18 cakes to be divided equally between her 2 shops. Copy and complete the expression below to find the number of doughnuts, tarts and cakes that she will deliver to each shop.

$$90d + 34t + 18c = 2(\qquad\qquad)$$

3 Roberto buys another disco and more records and tapes. He now has 560 singles (s), 100 LPs (*l*) and 35 tapes (*t*) to share between 5 discos. Use an expression like the ones above to find out the number of singles, LPs and tapes in each disco.

4 Senga is given another 2 newsagents to visit on her early morning delivery round. The 7 shops each get the same number of papers. Her totals are 175 *Guardians* (G), 294 *Records* (R) and 252 *Expresses* (E). Write out an expression to find out the number of *Guardians*, *Records* and *Expresses* she should deliver to each shop.

5 Shaheen is making up writing sets to sell on her stall at the local market. Each set contains pencils, a sharpener, a ruler and note pads. She has 150 pencils, 50 sharpeners, 50 rulers and 100 note pads. What is the maximum number of writing sets she can make up? Use shorthand and brackets to work out the number of pencils, rulers, sharpeners and note pads that should go into each set.

Factorise the following expressions. Make sure the term outside the brackets is the highest common factor.

6 $2x+4$	**7** $3a+15$	**8** $5+10x$
9 $3x+12y$	**10** $4p+16q$	**11** $7y+7z$
12 $5x-15$	**13** $54-9y$	**14** $6p-30$
15 $8x-64y$	**16** $3q-21r$	**17** $10r-10s$
18 $8x+12$	**19** $12p+15$	**20** $21+9r$
21 $14y+35z$	**22** $36s+27t$	**23** $15v+5w$
24 $5x+10y+15$	**25** $6a+3b+3$	**26** $12r+8s+20$
27 $9p+6q+12r$	**28** $10r+16s+6t$	**29** $5d+15e+25f$
30 $4c-6d+10$	**31** $15r+9s-21t$	**32** $50c-10d+35e$
33 $9p-18q-6$	**34** $18c-12d-30e$	**35** $14g-28h-7i$
36 $5x+xy$	**37** $pq+7p$	**38** $d+4c$
39 $st-8t$	**40** $rs-5s$	**41** $uv-4v$
42 $pq+pr$	**43** $cd-de$	**44** $xy+xz$
45 $ac+c$	**46** $p-pq$	**47** $t-st$
48 $3xy+6xz$	**49** $4c+6cd$	**50** $5pq+15qr$

Substitution

Information ➷ In maths **substitution** means replacing a letter in an expression with a number. The answer to the resulting calculation is then worked out. **B O D M A S** is one way to remember the order for calculations. It stands for **Bracket, Of, Divide, Multiply, Add, Subtract.** It tells you, for example, that you must multiply before adding or subtracting. For fractions, however, you must remember to work out the value of the top and bottom before dividing.

Example

Substitution problem

In Question 4 of Exercise 1, a shopkeeper checked the notes in each of her 3 tills. She worked out the following shorthand count for each till:

Till 1 $12f+4t$,   **Till 2** $11f+5t$ **Till 3** $14f+3t$

where f represents a £5 note and t a £10 note. How much money was in each till?

f stands for £5 and t stands for £10 so you begin by replacing f by 5 and t by 10:

	Till 1 $12f+4t$	**Till 2** $11f+5t$	**Till 3** $14f+3t$
Substitute f = 5, t = 10: Multiply: Add:	$12\times5+4\times10$ $60+40$ 100	$11\times5+5\times10$ $55+50$ 105	$14\times5+3\times10$ $70+30$ 100

The shopkeeper has £100 in till 1, £105 in till 2, and £100 in till 3.

Examples

Substitution in expressions

Let $a=2$, $b=5$, $c=6$ and $d=8$

1   Find the value of $b+c-a$

Substitute: $5+6-2 = 9$

2   Find the value of $4b-2a$

Substitute: $4\times5-2\times2$
Multiply then subtract: $20-4 = 16$

3   Find the value of $bc+ad$

Substitute: $5\times6+2\times8$
Multiply then add: $30+16 = 46$

4   Find the value of $\dfrac{a^2+c}{b}$

Substitute: $\dfrac{2^2+6}{5}$

Work out the top $\dfrac{10}{5} = 2$
line, then divide:

5   Use the same values of a, b, c, and d to make up expressions which equal 21.

$b+2d$ $3b+c$ $3a+3b$ b^2-2a

$5+2\times8 = 21$ $3\times5+6 = 21$ $3\times2+3\times5=21$ $5^2-2\times2 = 21$

Exercise 5

1   Sharon runs a class tuckshop selling crisps (11p), lemonade (16p) and biscuits (6p). On Monday she sold 10 bags of crisps, 15 bottles of lemonade and 8 biscuits. She uses the expression $10c + 15l + 8b$ to find her total sales, where c represents the cost of a packet of crisps. What do l and b represent? Substitute into the expression to find her total sales.

2 Bill has a stall selling electrical goods at the local market. On Saturday he sold 12 radios for £6 each and 35 cassette recorders for £18 each. He uses the expression 12r + 35c to find his total sales. What do *r* and *c* represent? Substitute into the expression to find his total sales.

3 Pete is organising a disco at his Youth Club. He sells single tickets for £1.50 each and double tickets for £2.90 each. He sells 40 single tickets and 35 double tickets. Copy and complete the expression for the total amount of money he receives. What do *s* and *d* represent? Substitute to find how much money he should have.

$$40s + \square d$$

4 Finlay is a farmer. At the monthly market he sold 30 sheep at £32 each and 10 cows at £65 each. Copy and complete the expression below to show how much money he receives. What do *s* and *c* represent? Substitute to calculate how much he receives.

$$\square s + \square c$$

5 One day Pauline sells 45 doughnuts at 16p each, 18 tarts at 45p each and 30 cakes at 40p each. Write down an expression to show how much she receives. Explain the meaning of any letters you use. Substitute to calculate the value of her sales.

If a = 1, b = 3, c = 4, d = 6, e = 7 and f = 8 find the values of

6 b+d	7 e−c	8 f−d	9 3f
10 9b	11 a+d−c	12 f−d+c	13 5c+2a
14 6d−8b	15 7b−f	16 5d+f	17 8b−3e
18 4f−5c	19 4d−5b+c	20 e−3c+f	21 5a−f+b
22 bd	23 df	24 bc−d	25 cd−ae
26 cf−2e	27 2bc	28 3ad	29 5be
30 bcd	31 abd	32 b^2	33 c^2
34 f^2	35 e^2+b	36 d^2	37 b^2-bc+c^2
38 $2b^2$	39 $5a^2$	40 $3c^2$	41 $\dfrac{bc}{a}$
42 abcd	43 a^2+b^2	44 $\dfrac{d^2+f}{c}$	45 $\dfrac{d^2-c}{f}$

Use the same values for *a, b, c, d, e,* and *f* to make at least 2 expressions to equal each of these numbers.

46 16   47 32   48 43

49 Use the same values of a, b, c, d, e, and f together with the numbers 1, 2, 3, 4, and 5 to make up as many expressions as possible to equal 55.

Using formulae

Information ↩ You have seen that a formula is a mathematical relationship. Once it has been established it can be used for many different examples of the same situation.

For example $a = lb$ is the formula used to find the area of a rectangle. We can use this formula to find the area of any rectangle.

When you have done the substitution you will need a calculator. Remember to make an estimate of the answer before using the calculator. This helps you to spot mistakes.

Example   **Evaluating a formula**

$R = C+P(T+W)$. Find R if $C=3.7$, $P=5.4$, $T=13.6$ and $W=5.8$

	$R = C+P(T+W)$
Substitute for C, P, T and W:	$R = 3.7+5.4\times(13.6+5.8)$

Make an estimate:	$R = 3.7+5.4\times(13.6+5.8)$
approximate each number	$4+5\times(14+6)$
do the bracket first	$4+5\times20$
multiply	$4+100$
add	104

Calculate the answer:	$R = 3.7+5.4\times(13.6+5.8)$
	$R = 3.7+5.4\times19.4$
	$R = 3.7 + 104.76$
	$R = 108.46$

The answer and the estimate are close enough to suggest that the calculation is correct.

Exercise 6   All these formulae are used in maths or science.

1 $V = IR$
Find V if $I=3.7$ and $R=54$

2 $F = ma$
Find F if $m=46.3$ and $a=6.4$

3 $C = \pi d$
Find C if $\pi=3.14$ and $d=7.8$

4 $P = I^2R$
Find P if $I=12.5$ and $R=2.1$

5 $A = \pi r^2$
Find A if $\pi=3.14$ and $r=17.5$

6 $A = \pi rs$
Find A if $\pi=3.14$, $r=15.7$ and $s=25.2$

7 $V = lbh$
Find V if $l=6.4$, $b=5.1$ and $h=2.7$

8 $E = mgh$
Find E if $m=4.5$, $g=9.8$ and $h=28$

9 $P = \dfrac{Fs}{t}$
Find P if $F=14.3$, $s=2.1$ and $t=3.5$

10 $E = \dfrac{mV^2}{2}$
Find E if $m=15.2$ and $V=4.3$

11 $W = \dfrac{mv}{g}$
Find W if $m=5$, $v=3.4$ and $g=9.8$

12 $h = \dfrac{gar}{v^2}$
Find h if $g=9.8$, $a=4.3$, $r=6.2$ and $v = 5$

13 $s = ut + \frac{1}{2}at^2$
Find s if $u=3$, $t=6.5$ and $a=14.3$

14 $V^2 = u^2 + 2as$
Find V^2 if $u=3.7$, $a=15.2$ and $s=9.3$
Can you find the value of V?

15 $E = mgh + \frac{1}{2}mV^2$
Find E if $m=6.4$, $g=9.8$, $h=0.5$ and $V=0.6$

16 $L = a+(n-1)d$
Find L if $a=16$, $n=7$ and $d=3$

17 $C = \dfrac{5(F-32)}{9}$
Find C if $F=68$

18 $T = \dfrac{12(D-d)}{L}$
Find T if $D=65.3$, $d=47.8$ and $L=103.4$

19 $F = \dfrac{W(v-u)}{gt}$
Find F if $W=53.2$, $v=19.6$, $u=12.7$, $g=9.8$ and $t=5.7$

20 $A = 2(lb+bh+lh)$
Find A if $l=23.4$, $b=18.5$ and $h=34.7$

21 $V = \pi r^2 h$
Find V if $\pi=3.14$, $r=16.5$ and $h=46.8$

22 $V = \dfrac{4\pi r^2}{3}$
Find V if $\pi=3.14$ and $r=12$

23 The extension of a spring is found by using the formula $e = 6.2l$, where e is the extension in cm and l is the load on the spring in kilograms. What will the extension be if a load of 8 kg is applied to the spring?

24 The time you would take to travel a certain distance is given by the formula

$$t = \frac{d}{s}$$

where t is the time you would take, d is the distance you have to travel and s is the speed you will be travelling at. How long will it take you to travel a distance of 375 km at a speed of 80 km per hour?

25 In electricity the power rating on a piece of equipment is found from the formula $P = VI$ where P is the power in watts, V is the voltage in volts, and I is the current in amps. Find the power rating on a fire that uses 6.25 amps at 240 volts.

26 In Physics the acceleration of a vehicle is found from the formula

$$a = \frac{v-u}{t}$$

where a is the acceleration in m/s², v is the final velocity in m/s, u is the initial velocity in m/s and t is the time in seconds. Find the acceleration of a vehicle if its speed increases from 3 m/s to 28 m/s in 4.75 seconds.

Constructing formulae

Information ⮎ Often the results of an investigation or an experiment in science are best described by using a formula. If you can find a formula it will often save you a lot of work.

Example **Constructing a formula**

The diagram shows 3 different polygons. In each polygon one vertex has been chosen and marked with a dot. All the diagonals from that vertex have been drawn. How many diagonals can be drawn from one vertex in a (a) 15-sided polygon (b) 50-sided polygon?

| 4 sides | 5 sides | 6 sides |
| 1 diagonal | 2 diagonals | 3 diagonals |

The table below shows the information from the polygons. This information can also be shown on a graph. Note that the points have not been joined up because you can only have a whole number of sides or diagonals.

Sides in the polygon (s)	Diagonals from 1 vertex (d)
4	1
5	2
6	3

The table can be extended to give

7	4
8	5
9	6
10	7
11	8
12	9
13	10
14	11
15	12

Part (a) of the problem can be solved by looking at either the graph or the table. The numbers in each column in the table follow a pattern. The points in the graph lie in a straight line. Look at the last point on the graph:

12 diagonals can be drawn from 1 vertex in a 15-sided polygon.

Extending the graph or table to solve part (b) would involve a lot of time and paper! Another approach is to look for a relationship or link between the number of sides and the number of diagonals. The relationship can be expressed as either a **word formula** or as a formula.

Word formula: The number of diagonals from 1 vertex is always 3 less than the number of sides.

Formula: $d = s-3$
where d is the number of diagonals from 1 vertex, and s is the number of sides in the polygon. (When you write down a formula it is important to state what each letter represents.)

We can use this formula to answer part (b). We substitute $s = 50$ into the formula.

$$d = s-3$$
Substitute: $\qquad d = 50-3$
Subtract: $\qquad d = 47$

In a 50-sided polygon 47 diagonals could be drawn from 1 vertex.

Investigations

1   Here are the 3 polygons again. In each one all the diagonals have been drawn from 1 chosen vertex. The polygons have been divided into triangles.

4 sides

2 triangles

5 sides

3 triangles

6 sides

4 triangles

No. of sides in polygon (s)	4	5	6	7	8	9	10
No. of triangles in polygon (t)	2	3	4				

(a)   Copy and complete the above table. Draw more polygons if necessary.

(b)   Draw a graph of s against t. Extend your graph to find the number of triangles in a 13-sided polygon.

(c)   Write down a word formula to describe the relationship between the number of sides and the number of triangles.

(d)   Copy and complete the formula:
$t = \ldots$ where t is the ... s is the ...

(e)   Use your formula to find the number of triangles in a 35-sided polygon.

2 The photograph shows Clifton Suspension Bridge in Bristol. The vertical supporting cables are 2 metres apart. Investigate the number of cables that are needed for different distances.

(a) Copy and complete this table.

Distance along bridge (d)	2 m	4 m	6 m	8 m	10 m
No. of cables (c)	1	2			

(b) Draw a graph of c against d.

(c) Write down a word formula explaining the relationship between the number of cables and the distance along the bridge.

(d) Copy and complete the formula connecting c and d.
c = . . . where c is the . . . d is the . . .

(e) The bridge on page 3 had supporting cables 5 m apart. Write a formula for that bridge.

(f) Rewrite the formula if the distance between the cables is x metres.

(g) The two towers of the Clifton Suspension Bridge are 90 m apart. Use your formula to find the number of cables between the towers.

3 Paul works for a mail order company that sells video tapes at £5 each plus a fixed charge of £2 for post and packing (P&P). He needs to know how much to charge for any number of video tapes.

(a) Copy and complete this table.

No. of video cassettes ordered (N)	1	2	3	4	5
Total cost (inc. P & P) (C)					

(b) Draw a graph of C against N.

(c) Write down a word formula and formula connecting C and N.

(d) Use your formula to find the cost of buying 25 video cassettes.

(e) If post and packing are increased to £3, rewrite your formula.

(f) Rewrite the formula if post and packing is £P.

4 A recipe for roast beef gives the following instructions for finding the cooking time.

'Allow 20 minutes per pound weight plus an additional 20 minutes'

(a) Copy and complete the table below.

Weight of meat in lbs (W)	1	2	3	4	5	6
Cooking time in hours (t)						

(b) Draw a graph of t against W.

(c) Extend the graph to find the cooking time for a 9lb roast.

(d) Write down a word formula and formula connecting t and W.

(e) Use the formula to find the cooking time for a 4½ lb roast.

5 A car hire firm charge £25 plus £0.04 per mile to hire a car for 1 day.

(a) Copy and complete this table.

Distance travelled in miles (d)	100	200	300	400	500
Total hire charge (c)					

(b) Draw a graph of c against d.

(c) From the graph find the hire charge for a motorist who travels 600 miles.

(d) Write down a word formula and formula connecting c and d.

(e) Use the formula to calculate the hire charge for a motorist who drives 850 miles.

(f) Write down a formula to calculate the total cost of hiring a car for t days if the charges are £25 per day plus £0.04 per mile.

(g) Use your formula to calculate the total cost of hiring a car for 5 days and travelling 1300 miles.

Camp stool puzzle

Look back at Cuong's puzzle. Continue the sequence to give you more data.

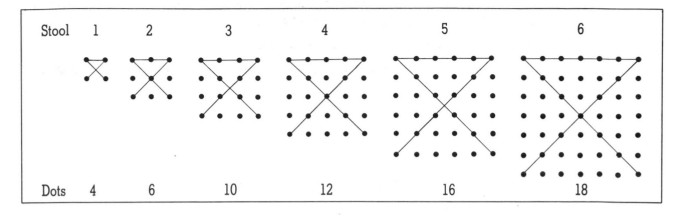

Now make up a table.

Stool (s)	No. of dots (d)	Difference
1	4	
2	6	2
3	10	4
4	12	2
5	16	4
6	18	2

The difference suggests that two separate tables are needed.

Let's try one for even stool numbers and one for odd stool numbers:

Table (even)

Stool (s)	Dots (d)
2	6
4	12
6	18

Word formula: Number of dots is 3 times stool number.

Formula: $d = 3s$

Table (odd)

Stool (s)	Dots (d)	3s
1	4	3
3	10	9
5	16	15

Word formula: Number of dots is 3 times stool number plus 1

Formula: $d = 3s + 1$

The answers Michelle gave are now easy to calculate.

For 24th Stool (even)
Use $d = 3s$ where $s = 24$
$= 3 \times 24$
$= 72$ dots

For 15th stool (odd)
Use $d = 3s + 1$ where $s = 15$
$= 3 \times 15 + 1$
$= 46$ dots

Can you see why there is a split solution for odd and even numbers? (Hint: look at the centres of the stools.)

Dotty investigations

Investigations   Here are 3 dotty investigations. In each case you have to find the relationship between the shape number and the number of dots in the shape. The first 3 shapes are given in each pattern. The following instructions may help.

(a) Draw out the first 5 shapes in each pattern.

(b) Draw a table headed Shape Number (s), No. of dots (d). Part of the table is shown for investigation 1.

(c) Draw a graph to check that all 5 points lie in a straight line.

(d) Write out a word formula.

(e) Write out a formula.

(f) How many dots are there in the 10th and 25th shapes?

1
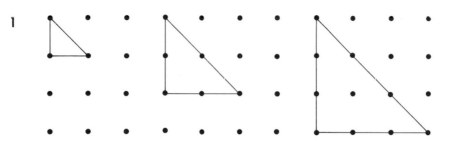

Shape number (s)	Number of dots (d)
1	3
2	6
3	9

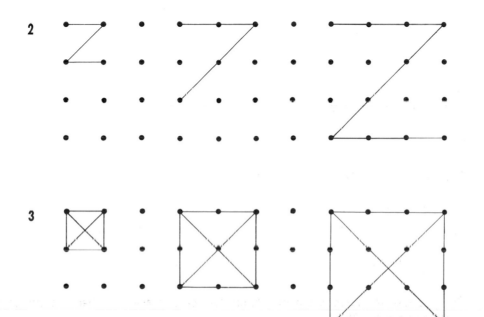

4 Navigation

Do you remember? ➷

In this unit it is assumed that you know how to:

1 use a coordinate grid

2 round answers

3 use the metric system.

Exercise 0 will help you check. Can you answer all the questions?

Exercise 0

Using coordinates

1 Below is a grid containing 25 squares. All the letters of the alphabet, except 'Z', are arranged in a random way.

Use the grid to write down the message given by these coordinates. (5, 1) Refers to **H** not **A**.
(5,1) (1,5) (5,4) (3,1)
(1,5)
(5,1) (1,5) (2,3) (2,3) (4,1)
(2,2) (1,5) (4,1)

5	A	R	J	T	I
4	Q	B	S	U	V
3	C	P	K	G	O
2	L	D	W	F	X
1	M	N	E	Y	H
0	1	2	3	4	5

2 Write down the coordinates for the message 'NO SMOKING'

Rounding

Round these answers to 2 significant figures.

3 23.783

4 3.825

5 5.3792

6 0.72256

7 0.6853

The metric system

Change each of these measurements to the units stated.

8 3700 cm to metres

9 45 000 mm to metres

10 416 500 cm to metres

11 56 300 m to kilometres

12 256 000 cm to kilometres

13 4 560 000 mm to kilometres

Investigation

The knights tour

The squares on chess boards can be identified by using the letters A to H and numbers 1 to 8 as shown.

The knight moves in an 'L' shape, 2 squares in 1 direction then 1 square at right angles. Ask your teacher to explain how a knight moves in chess if you are unsure.

Starting with the knight on square (A,1) can you plot a path so that the knight lands on each square only once? Record your path using coordinates.

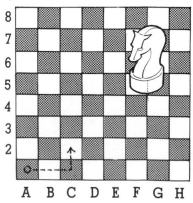

The best route

Mr and Mrs Kerray are driving from Edinburgh to York for a week's holiday. Their 2 teenage children, Pauline and Darren are planning a trip to the Lake District the same week.

Mr and Mrs Kerray want to drop Pauline and Darren in Keswick on their way to York. They need to choose a route which takes them from Edinburgh to Keswick to York. Can you help them? They also need to know roughly how far they will be travelling, in total, so that they can have some idea how long the trip will take them.

Edinburgh Castle

York Minster

After you have worked through this unit you will be able to help them to work out the best route they could take, and work out the distance from road maps.

The national grid

Information ⇨ Below is a map of Britain showing the **National Grid**. Each grid square is 100 km by 100 km. A square is identified by giving the coordinates of the bottom left (South-West) corner. No bracket or comma is used between the coordinates of a map reference.

Example **Map references from the National Grid**
What is the reference for the square containing Aberdeen?

The horizontal coordinate of this square is 300
The vertical coordinate of this square is 700
The map reference is 300700

Exercise 1 Use the map below to find the map reference for the square containing:

1 Norwich **2** Cardiff **3** Newcastle **4** Edinburgh **5** Manchester **6** Southampton

Which city is in square

7 200600 **8** 300200 **9** 500100 **10** 400300 **11** 200000 **12** 400400

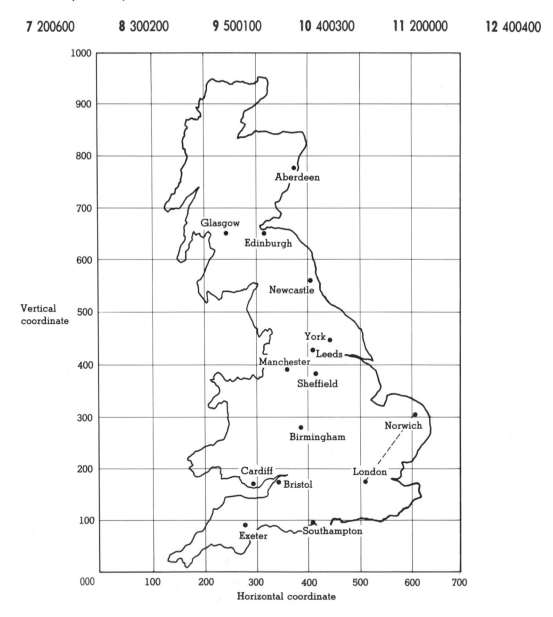

Map references

Information ➥ Each 100 km square of the National Grid may contain many towns and villages. The map references you have been using are seldom accurate enough to find all these places on the map so each 100 km square is divided up into 10 km then 1 km squares depending on the accuracy needed. To identify a place by its square you should give the coordinates of the bottom left (South-West) corner of the square. Each coordinate will have 2 figures making a **4 figure map reference**.

Example

4 figure references
Mrs Kerray has bought a map of the York area and a guide book which gives map references. The map on this page is a map of grid square 400 400 on the national grid divided into 10 km squares which are drawn and numbered. The map shows the main roads, towns, villages and some places of interest. What is the the 4 figure map reference for York?

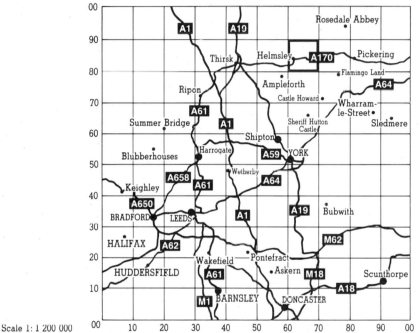

Scale 1: 1 200 000

The horizontal coordinate is 60, the vertical coordinate is 50
The 4 figure reference is 6050

Exercise 2

1 Use the map to write down the name of the village or villages in the squares with grid references
   (a) 1050   (b) 7030   (c) 8060   (d) 9060

2 Find the following places and give their 4 figure map reference.
   (a) Ripon   (b) Shipton   (c) Ampleforth   (d) Askern

3 Here is a list of the places Mr and Mrs Kerray want to visit.

(a) Flamingo Land 7070   (b) Pickering Castle 8080   (c) Sherrif Hutton Castle 6060

(d) Castle Howard 7070   (e) Rosedale Abbey 7090   (f) Wetherby Race Course 4040

Find each of these places on the map. List the places that are close to each other and which could be visited on the same day.

Information ➭   The map on this page is a larger scale map of grid square 6080 in the previous map. It is divided into 1 km squares. You can still use a 4 figure map reference to identify a place on this map but sometimes you need to be more accurate and use a **6 figure map reference.**

Scale 1:60 000

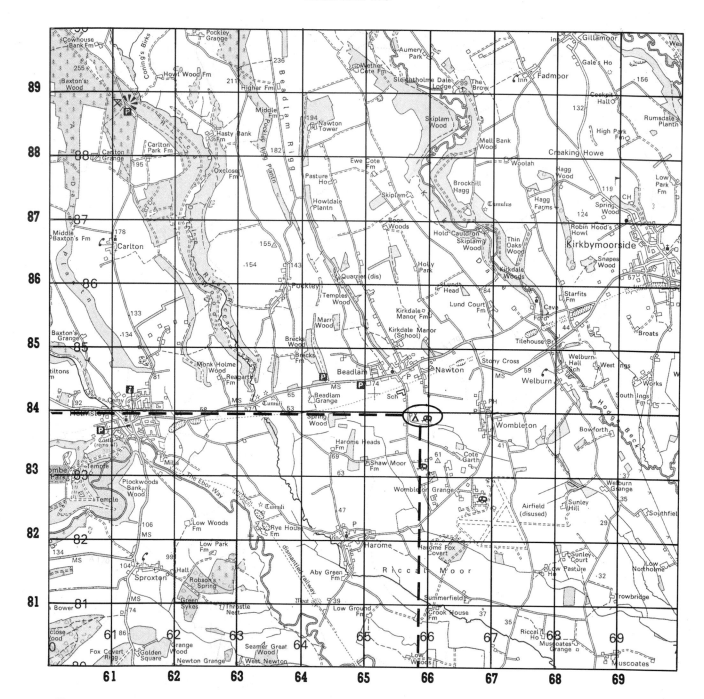

Example   **Identifying a place given its 6 figure reference**
What is at 658839?

658 tells you that the place is 0.8 km east of line 65

839 tells you that it is 0.9 km north of line 83

Looking at the map, at this reference there is a camping and caravan site.

Example

Giving a 6 figure map reference
Find the 6 figure reference for Low Park Farm. Low Park Farm is in square 6281.

Imagine this 1 km square is divided into one hundred 100 m squares as shown.
Low Park Farm is approximately **0.6** km East and **0.8** km North of the South West corner of 6281.

The 6 figure reference is 626818.

This defines a 100 m square as shown.

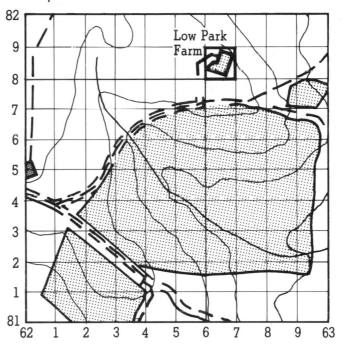

Exercise 3

Use the map on page 61 to identify these places from the 6 figure map given.

1 644834	2 681825	3 652869	4 622822
5 689889	6 644876	7 689837	8 641884
9 656875	10 650831	11 675894	12 682847

Using the same map, give the 6 figure map reference for the following

13 Inn in square 6789

14 Boon Woods in square 6586

15 Starfits Farm in 6885

16 Middle Farm in 6388

17 Castle in 6183

18 Church in 6186

19 Higher Farm in 6389

20 Trig Point in 6386

21 Broats in 6985

22 Crook House Farm in 6680

23 Rye House Farm in 6382

24 Wombleton Grange in 6682

25 The Tourist Information Office at Helmsley

26 The Church in 6186

27 The caravan site in 6682

28 The church in 6482

29 The phonebox in 6784

30 The school in 6585

Distances from maps

Information ➪ A map is a **scale drawing** of an area of ground. If you want to change the measurements from a map to the actual size you should **multiply by the scale factor**. It is important to express the answer in the most appropriate units.

Example

Scaling up measurements from maps
Raschid is going on a cycle trip. He has bought a map with a scale of 1:250 000. A path on the map is 3.4 cm long. How long is the actual path ?

$$3.4 \text{ cm} \times 250\ 000 = 850\ 000 \text{ cm}$$
$$850\ 000 \text{ cm} = 8500 \text{ m}$$
$$8500 \text{ m} = 8.5 \text{ km}$$

Exercise 4

Copy and complete this table.

	Measurement from map	Scale of map	Full size measurement
	3.4 cm	1:250 000	8.5 km
	5.2 mm	1:50 000	260 m
1	3.6 cm	1:250 000	
2	6.5 cm	1:50 000	
3	8 mm	1:500 000	
4	9.5 mm	1:100 000	
5	14.3 cm	1:100 000	
6	7.6 mm	1:50 000	

7 On a map of scale 1:500 000 the distance between 2 towns is 6.3 cm. How many kilometres is it between the 2 towns ?

8 On a map of scale 1:400 000 a section of motorway is 28 mm long. How long is the motorway ?

9 Angela is going hill walking in the Peak District. She has a map of the area drawn to a scale of 1:50 000. On the map, 2 hills are 24 mm apart. How far will she have to walk between the hills ?

10 A golf course occupies a rectangular piece of land 2 km by 3 km. The club committee want a new map of the course drawn to fit onto A4 paper. What scale should they use for the map?

Information ⟿ You can now use maps to calculate the distance between any two places provided you know the scale factor. Scale factors are always given on the map. The example and exercise below refer to the map on page 60.

Example **Calculating actual distances from a map**
Use the map on page 60 to find the distance from York to Leeds.

From the map York to Leeds is 3 cm. The scale of the map is 1: 1 200 000
Distance 3 cm × 1 200 000 = 3 600 000 cm to 2 significant figures
3 600 000 cm = 36 000 m
36 000 m = 36 km

Exercise 5 Use the map on page 60 to find the actual distance between the following:

1 York and Harrogate
2 York and Doncaster
3 Bradford and Leeds
4 Barnsley and Harrogate
5 Leeds and Pontefract
6 Scunthorpe and York
7 Doncaster and Thirsk
8 Harrogate and Ripon
9 Thirsk and York
10 Thirsk and Ripon
11 Barnsley and York
12 Scunthorpe and Harrogate

The distances you have been measuring are the straight line distances from place to place. Roads do not often run in straight lines and the actual distance to be driven or cycled between all the pairs of towns will be more than you have calculated.

13 Look carefully at the map on page 60. For questions **7** to **12** above, estimate the distance of the journey by road. Explain how you estimated the distance by road.

14 Mr and Mrs Kerray plan to visit friends in Harrogate and then to travel to Pontefract before returning to York. Estimate the distance they will travel by road.

Investigation **Mapping**
Find the area of land occupied by your school. Make a map of your school, choosing a scale so that the map will fit onto A4 paper. Mark on the map the position of any football, hockey or rugby pitches, tennis courts, playgrounds and all buildings.

The compass

Information ⇨ For hundreds of years travellers have used the **mariner's compass** or a similar instrument to find their way across long and difficult journeys.

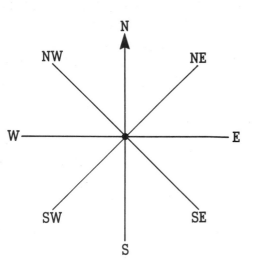

The diagram shows a simple **8 point compass**. It is divided into the 4 main directions: North, South, East and West and 4 intermediate directions: North East, South East, South West and North West.

Example

Estimating direction
Here is a map of England and Wales with some cities marked. The vertical lines all run North-South and the horizontal lines all run East-West. Estimate the direction from London to Norwich.

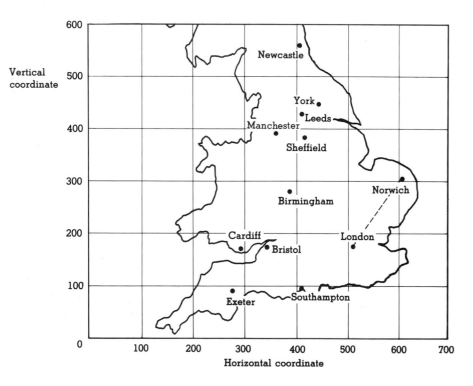

Norwich is approximately North East (NE) of London.

Exercise 6 Find the direction from the first to the second town:

1 London to Bristol   2 London to Birmingham   3 Leeds to Newcastle

4 Leeds to Sheffield   5 Leeds to Manchester   6 Southampton to Bristol

7 Southampton to Exeter   8 Southampton to Sheffield   9 Newcastle to Southampton

Information ⇨ Directions can be given more accurately if the compass is divided into 16 points as shown opposite.

When naming the 8 new points you always start with one of the 4 main points — North, South, East or West.

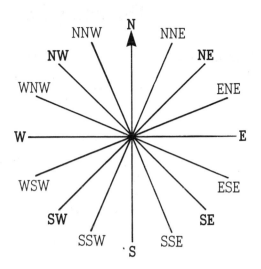

Example

Naming compass points
Name the compass direction which is midway between South West and West.

Start with West then add South West to give West South West (WSW).

Exercise 7

1 Write out in full NNE, ESE, SSE and NNW.

2 Copy out the compass shown above and mark in all 16 directions.

3 Use the 16-point compass and the map of England and Wales on the previous page to complete the table.

	From	To	Direction
	London	Sheffield	NNW
(a)	London	Exeter	
(b)	London	Birmingham	
(c)	Newcastle	York	
(d)	Birmingham	Manchester	
(e)	Birmingham	Sheffield	
(f)	Bristol	Birmingham	
(g)	Cardiff	Norwich	
(h)	Cardiff	Manchester	
(i)	Newcastle	Manchester	

Investigation

Opposite directions
In what direction would I be travelling if I went from Cardiff to Birmingham?
In what direction would I be travelling if I went from Birmingham to Cardiff?
What do you notice about the 2 answers?
Is this pattern the same for all journeys?
Choose 5 more pairs of cities shown on the map and work out the compass direction for each journey.
Write down your conclusions.

Bearings

Information ↝ With only 16 points on a compass the names given to the directions are already getting long and complicated. An easier way to describe directions accurately is to use **bearings**. A bearing is an angle measured clockwise from North.

There are 360° in 1 revolution. Three figures are always used to give a compass bearing in degrees. The 8 point mariners compass is shown opposite.

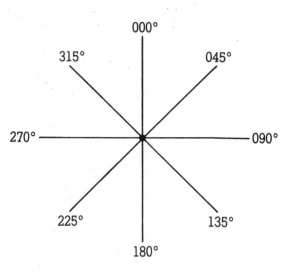

Example

Using a 360° protractor to find a bearing
Use the map of Britain showing the National Grid given on page 59 to find the bearing of London from Edinburgh.

1   Place centre of protractor on Edinburgh

2   Make sure 0° is pointing North

3   Read off bearing at London
The bearing from Edinburgh to London is 156°.

If you only have a 180° protractor your teacher will explain how to measure bearings greater than 180°.

Exercise 8

Use a 360° protractor and the map on page 59 to complete the table.

	From	To	Bearing
	Glasgow	Edinburgh	085°
	Glasgow	Newcastle	120°
1	Glasgow	Bristol	
2	Glasgow	Southampton	
3	Edinburgh	Newcastle	
4	Edinburgh	Leeds	
5	Edinburgh	Cardiff	
6	Edinburgh	London	
7	Norwich	Manchester	
8	Birmingham	Norwich	
9	London	Bristol	
10	Southampton	Leeds	

Investigation

Back Bearings
You have already investigated opposite directions but what happens to bearings if you reverse the journey?
Choose 5 of the bearings you have found in exercise 8, and work out the bearing for the opposite journey.
Write down a rule which allows you to calculate this bearing which is called the **back bearing**.

Information ⇨ The use of scales and bearings help to solve many navigational problems. Radar is used to find distances and bearings.

Example

Bearings and distances from radar screens
The diagram shows the situation that might be seen on a radar screen at Heathrow Airport. The distance between each circle represents a distance of 5 km. The airport is at the centre of the screen. Aeroplane A is 5 km from the airport on a bearing of 060°

Scale: 1 cm represents 5 km

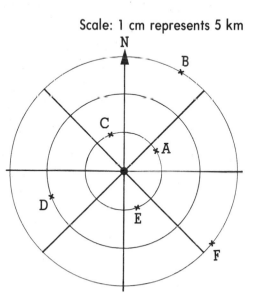

Exercise 9

1 Write down the direction and distance of the other planes from Heathrow.

2 This radar screen is in an Air-Sea rescue helicopter which is hovering above a damaged ship. The blips are other rescue ships which are in the area.

How far, and in which directions should the pilot tell the rescue ships to sail to get to the stricken vessel?

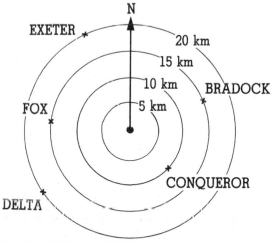

Navigation problems

Information ⇨   Making a scale drawing of a problem helps you to work out distances and directions accurately.

Example   **Scale drawing to solve a navigation problem**
Pauline and Darren have decided to hire a boat for a day on Lake Windermere. They sail for 8 km on a bearing of 060° then for 5 km on a bearing of 210°. How far, and in which direction should they go to get back to their starting point?

Choose a suitable scale.

1 cm represents 1 km or 1:100 000
Make an accurate scale drawing.

Measure the bearing and the distance back to the starting point.
Use the scale factor to find the actual distance.

Pauline and Darren will get back to the starting point if they sail for 4.5 km on a bearing of 275°.

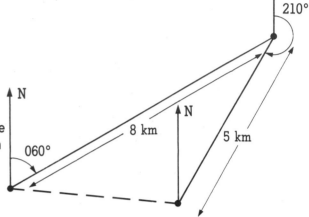

Exercise 10   The map shows Lake Windermere where Pauline and Darren spend some time boating. The map is reproduced here to a scale of 1:25 000

Use the map to measure the distances and bearings they sail for each of the trips below. Draw an accurate sketch of each trip.

1   Cockshot Point to The Ferry House.

2   The Ferry House to Ellerwood.

3   Ellerwood to Cockshot Point.

Scale 1:25 000

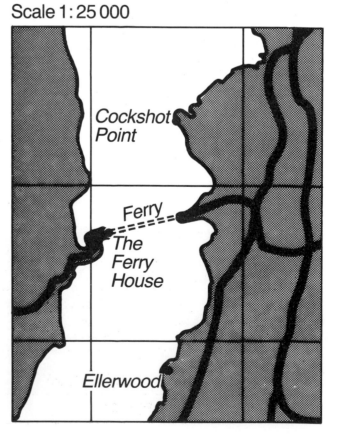

Hill walking

Information ☞ Pauline and Darren want to do some hill walking in the Lake District. For each walk which they want to do they make up 2 **route cards**. One copy they take with them and the other they leave with the police in case of an accident. They use what is known as Naismith's Rule to calculate the times for the card:

Allow 1 hour for every 4 km walked and 1 min for every 10 m climbed

Example

Making a route card

Here is Pauline and Darren's card for the first walk they intend to do.

ROUTE CARD

From	To	Distance	Height	Time
Camp Site Scafell Pikes Lingmell Crag	Scafell Pikes Lingmell Crag Camp Site	4 Km 1 Km 3 Km	900 m 80 m	1hr + 90 min = 2:30 15min + 8min = :23 45min = :45
TOTALS Add 10 mins per hour (or part)		8 Km	980m	3:38 :40
TOTAL TIME FOR TRIP				4:18

Exercise 11

Make some blank route cards like the one in the example. Copy the following routes onto your blank cards and complete them to find the total time that should be set aside for each trip.

ROUTE CARD 1

From	To	Distance	Height	Time
Road End Little Man Skiddaw	Little Man Skiddaw Road End	3 Km 1½ Km 5 Km	670 m 70 m	
TOTALS Add 10 mins per hour (or part)				
TOTAL TIME FOR TRIP				

ROUTE CARD 2

From	To	Distance	Height	Time
Camp site Black Sail Pass Pillar	Black Sail Pass Pillar Road End	5 Km 2½ Km 5 Km	460m 360m	
TOTALS Add 10 mins per hour (or part)				
TOTAL TIME FOR TRIP				

The best route

Here is one possible route the Kerrays could take.

Get a road map for example, AA road map of Britain, and follow the route.

From	To	Road	Direction	Distance
Edinburgh	Galashiels	A7	SE	50 km
Galashiels	Hawick	A7	S	}
Hawick	Langholm	A7	S	} .... 96 km
Langholm	Carlisle	A7	S	}
Carlisle	Bothel	A595	SW	}
Bothel	Keswick	A595	SE	} .... 50 km
Keswick	Penrith	A66	E	}
Penrith	Greta Bridge	A66	E	} .... 112 km
Greta Bridge	Scotch Corner	A66	SE	}
Scotch Corner	Boroughbridge	A1	SE	}
Boroughbridge	York	A59	E	} .... 64 km

Total distance is approximately 372 km. This was measured by moving a ruler along a map. Can you think of a more accurate method of finding the distance?

Choose another route and describe it in a table like the one above.

List the things you must think about when choosing a route.

Use a road map to estimate the total length of this route.

From	To	Road	Direction	Distance
Edinburgh	Biggar	A702	SW	
Biggar	Crawford	A702	SW	
Crawford	Carlisle	A74	S	
Carlisle	Penrith	M6	S	
Penrith	Keswick	A66	W	
Keswick	Kendall	A591	SE	
Kendall	Skipton	A65	SE	
Skipton	Harrogate	A59	E	
Harrogate	York	A59	E	

Investigations

1   Plan a route between the following pairs of cities. Describe the route in a table and a sketch.

    (a) Bristol to York   **(b)** Liverpool to Dover   **(c)** Norwich to Exeter

2   **(a)** Get a map of your local area. In towns you can use a street map. Make up some route cards for trips round your local area.

    (b) Try making up route cards which will give routes lasting for a certain length of time for example 2 hours.

    (c) If this is done by the whole class you will be able to build up a bank of local walks.

    (d) If you try the walks, write down anything interesting that others should look out for. Pool this information for the class.

3   Sheila and Fatima like hill walking. They are careful to plan their routes. They use Naismith's rule to calculate the time each part of the route should take. Unfortunately they never manage to keep to these timings and find that they are slower than their calculations allow.

    Here are 2 examples of the actual times they took.
    6 km and 300 m of climbing took 2 hours and 30 minutes.
    9 km and 200 m of climbing took 3 hours and 20 minutes.

    Can you discover the variation of Naismith's rule that they should use to give them correct timings?

4   You will need a compass, preferably a Silva type. Ask your teacher to show you how to 'walk on a bearing'.
    Go into the middle of the playground. Mark your starting point then follow this course.

    20 paces 056°, 20 paces 146°, 20 paces 236°, 20 paces 326°.
    Where do you finish?

    Try this course.

    20 paces 138°, 20 paces 210°, 20 paces 282°, 20 paces 354°, 20 paces 066°.
    Where do you finish?

    Try this course.

    15 paces 208°, 15 paces 268°, 15 paces 328°, 15 paces 028°, 15 paces 088°, 15 paces 148°, 15 paces 208°.
    Where do you finish?

    Can you explain why the result is the same in every case?
    Does the starting point matter?
    Make up more courses like the ones above with 8 legs. Do they work?

5 *Equations*

Do you
remember? ⇨ In this unit it is assumed that you know how to:

1 substitute numbers into a formula

2 find the inverse process

3 solve simple equations

4 plot points on a graph.

Exercise 0 will help you check. Can you answer all the questions?

Exercise 0
Substitution
If $a=2$, $b=3$ and $c=1$ find the value of

1 $a+b$	2 $a+c$	3 $b+c$
4 $b-a$	5 $b-c$	6 $2a$
7 $3c$	8 $2a+b$	9 $3c-b$
10 $4a-b$	11 $2a+2b$	12 $3b+2c$
13 $5c-2a$	14 $2b-3c$	15 $a+2b-c$

Inverse processes
Write down the opposite (or inverse process) to

16 adding 5	17 subtracting 3	18 multiplying by 4
19 dividing by 2	20 dividing by 6	21 adding 4
22 subtracting 10	23 multiplying by 7	24 $+8$
25 -12	26 $\times 9$	27 $\div 7$
28 $+19$	29 $\times 7$	30 $\div 12$

Simple equations
Find the number to replace the ? in each of these equations so that the sum is correct.

31 $?+2 = 5$	32 $?+3 = 7$	33 $?+9 = 15$
34 $?-5 = 8$	35 $?-2 = 10$	36 $?-15 = 7$
37 $2\times? = 10$	38 $3\times? = 12$	39 $4\times? = 20$

Plotting points

40 Plot the points (1,3) (2,4) (3,5) (4,6). Join them up. What do you get?

41 Plot the points (0,1) (1,3) (2,5) (3,7). Join them up. What do you get?

42 Plot the points (0,5) (1,4) (2,3) (3,2) (4,1) (5,0). Join them up. What do you get?

At the circus

Alison and Niven have a young daughter. They decide to take her to the circus and arrange to meet Pauline, Mike and their 3 sons there.

Alison and Niven are first to arrive and buy all 8 tickets costing £12. Alison does not know how much Pauline owes her for her family's 5 tickets. Niven notices that the next lady in the queue buys 2 adult and 4 child's tickets and is charged £8.

Niven says he can work out how much Pauline owes Alison. Can you see how?

This unit explains about equations, graphs, simultaneous equations, inequations and will show you how to solve problems like this one.

Processes

Information ⌐ A **process** changes one number into another. It is a series of instructions like 'add 3' or 'multiply by 5'. A process is often made clearer by a **process diagram**. Each instruction is shown by a box, so that for 'add 3' you write $\boxed{+3}$, for 'multiply by 5' you write $\boxed{\times 5}$.

Examples

Changing one number into another
Find the number to replace the ? in each process diagram.

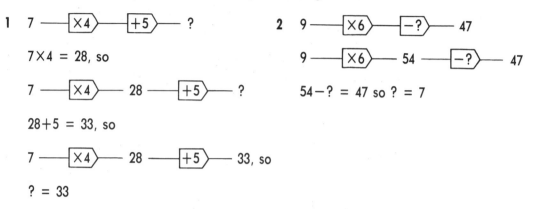

1   7 ——$\boxed{\times 4}$——$\boxed{+5}$—— ?

    $7 \times 4 = 28$, so

    7 ——$\boxed{\times 4}$—— 28 ——$\boxed{+5}$—— ?

    $28 + 5 = 33$, so

    7 ——$\boxed{\times 4}$—— 28 ——$\boxed{+5}$—— 33, so

    ? = 33

2   9 ——$\boxed{\times 6}$——$\boxed{-?}$—— 47

    9 ——$\boxed{\times 6}$—— 54 ——$\boxed{-?}$—— 47

    $54 - ? = 47$ so ? = 7

Information ⌐ A process can also be used to change a letter into an expression. Remember that letters (like x) are used to stand for any number.

Example

Drawing a process diagram for any number
Write out the process diagram for

1   7 times a number plus 5.   2   3 times a number take away 5.

The process can begin with any number. Use x to stand for any number.

x ——$\boxed{\times 7}$—— $7x$ ——$\boxed{+5}$—— $7x+5$ x ——$\boxed{\times 3}$—— $3x$ ——$\boxed{-5}$—— $3x-5$

Exercise 1

Find the missing number in the processes in questions **1** to **10**.

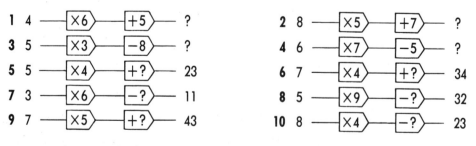

1   4 ——$\boxed{\times 6}$——$\boxed{+5}$—— ?   2   8 ——$\boxed{\times 5}$——$\boxed{+7}$—— ?

3   5 ——$\boxed{\times 3}$——$\boxed{-8}$—— ?   4   6 ——$\boxed{\times 7}$——$\boxed{-5}$—— ?

5   5 ——$\boxed{\times 4}$——$\boxed{+?}$—— 23   6   7 ——$\boxed{\times 4}$——$\boxed{+?}$—— 34

7   3 ——$\boxed{\times 6}$——$\boxed{-?}$—— 11   8   5 ——$\boxed{\times 9}$——$\boxed{-?}$—— 32

9   7 ——$\boxed{\times 5}$——$\boxed{+?}$—— 43   10   8 ——$\boxed{\times 4}$——$\boxed{-?}$—— 23

Write out the process diagram for each of the processes in questions **11-20**. Use x to stand for any number.

11   5 times a number plus 8   12   3 times a number plus 5

13   7 times a number plus 4   14   6 times a number plus 5

15   4 times a number take away 3   16   2 times a number take away 5

17   9 times a number take away 4   18   3 times a number take away 8

19   6 times a number plus 7   20   7 times a number take away 1

Information ⮐ The **inverse process** is the opposite process to the one given. The given process changes one number into another, or changes a letter into an expression. The inverse process takes you back to the original number, or takes you back from an expression to the original letter. To draw the inverse process diagram, you do the opposite of each step in the given process and draw the steps in the opposite order.

Examples

Finding the inverse process
Write out the inverse process for each diagram

1   6 ——⟨×7⟩—— 42 ——⟨+5⟩—— 47

The opposite of ×7 is ÷7 and the opposite of +5 is −5. You do these steps in the opposite order. So the inverse process is

47 ——⟨−5⟩—— 42 ——⟨÷7⟩—— 6

2   8 ——⟨×3⟩—— 24 ——⟨−7⟩—— 17

The opposite of ×3 is ÷3 and the opposite of −7 is +7. Do these steps in the opposite order. So the inverse process is

17 ——⟨+7⟩—— 24 ——⟨÷3⟩—— 8

Example

Finding the inverse process for any number
Write down the process and inverse process for 7 times a number plus 5.

Use x to stand for any number.

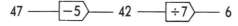

Process: x ——⟨×7⟩—— $7x$ ——⟨+5⟩—— $7x+5$

Inverse process: $7x+5$ ——⟨−5⟩—— $7x$ ——⟨÷7⟩—— x

Exercise 2

Copy and complete each of the processes in questions **1** to **8**. Below each write out the inverse process. Check that the inverse process takes you back to the original number.

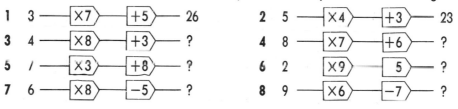

1   3 ——⟨×7⟩—⟨+5⟩—— 26   2   5 ——⟨×4⟩—⟨+3⟩—— 23
3   4 ——⟨×8⟩—⟨+3⟩—— ?   4   8 ——⟨×7⟩—⟨+6⟩—— ?
5   1 ——⟨×3⟩—⟨+8⟩—— ?   6   2 —⟨×9⟩——⟨5⟩—— ?
7   6 ——⟨×8⟩—⟨−5⟩—— ?   8   9 ——⟨×6⟩—⟨−7⟩—— ?

Write out the process and the inverse process for questions **9-16**. Use x to stand for any number.

9 2 times a number plus 5   10 6 times a number plus 1

11 3 times a number plus 4   12 7 times a number plus 6

13 4 times a number take away 5   14 6 times a number take away 3

15 8 times a number take away 7   16 3 times a number take away 7

Information ⇝ You have already seen how the inverse process reverses the process. This gives a method of finding the starting number if the end number is known.

Example   **Using the inverse process**
7 times a number plus 4 equals 25. What is the number?

Use x to stand for the unknown number.

Process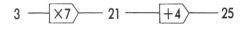

Inverse process $7x+4$ ─── ⟨−4⟩ ─── $7x$ ─── ⟨÷7⟩ ─── x

Use 25 in the inverse process.

25 ─── ⟨−4⟩ ─── 21 ─── ⟨÷7⟩ ─── 3

The number is 3. Check by starting the process with 3.

3 ─── ⟨×7⟩ ─── 21 ─── ⟨+4⟩ ─── 25

Exercise 3   Solve each of the following problems by

(a) drawing a diagram to show the process

(b) drawing a diagram to show the inverse process

(c) using the inverse process to solve the problem.

Check each answer as shown in the example above.

1 3 times a number plus 5 equals 17. What is the number?

2 4 times a number plus 7 equals 27. What is the number?

3 5 times a number plus 8 equals 18. What is the number?

4 9 times a number plus 5 equals 68. What is the number?

5 3 times a number take away 4 equals 17. What is the number?

6 7 times a number take away 8 equals 13. What is the number?

7 6 times a number take away 5 equals 31. What is the number?

8 8 times a number take away 7 equals 33. What is the number?

9 When a certain number is multiplied by 6 and then 5 is added the answer is 47. What is the number?

10 A certain number is multiplied by 5 and then 8 is taken away. The answer is 37. What is the number?

11 A number is multiplied by 15 and 17 is added. The answer is 197. What is the number?

12 A certain number is multiplied by 25 and 73 is taken away. The answer is 377. What is the number?

Solving equations

Information ➘ You have been using the process and inverse process to find an unknown number. These problems can also be solved using **equations**. A simple equation can be shown as a process diagram in which a letter, usually x, stands for an unknown number. The process gives the equation and the inverse process tells you how to find the unknown number in the equation. Remember that an equation must always balance. Whatever you do to one side must also be done to the other.

Examples

Finding the unknown number in an equation

1 7 times a number plus 4 equals 25. What is the number?

Use x to stand for the number.

Process: x ———$\boxed{\times 7}$—— 7x ——$\boxed{+4}$—— 7x+4

Equation: $7x+4 = 25$

The inverse process tells you how to find the value of x. If you 'do' each instruction in the inverse process to each side of the equation, the left hand side becomes just x and the right hand side gives you the value of x.

Inverse process:	7x+4 ——$\boxed{-4}$—— 7x ——$\boxed{\div 7}$—— x
Equation:	$7x+4 = 25$
Do $\boxed{-4}$ to each side:	$\boxed{-4}$ $\boxed{-4}$
	$7x = 21$
Do $\boxed{\div 7}$ to each side:	$\boxed{\div 7}$ $\boxed{\div 7}$
	$x = 3$

Check by starting the process with 3:

3 ——$\boxed{\times 7}$—— 21 ——$\boxed{+4}$—— 25

2 5 times a number take away 7 equals 28. What is the number?

Use x to stand for the unknown number.

Process:	x ——$\boxed{\times 5}$—— 5x ——$\boxed{-7}$—— 5x−7
Equation:	$5x-7 = 28$
Inverse process:	5x−7 ——$\boxed{+7}$—— 5x ——$\boxed{\div 5}$—— x
Equation:	$5x-7 = 28$
Do $\boxed{+7}$ to each side:	$\boxed{+7}$ $\boxed{+7}$
	$5x = 35$
Do $\boxed{\div 5}$ to each side:	$\boxed{\div 5}$ $\boxed{\div 5}$
	$x = 7$

Check by starting the process with 7:

7 ——$\boxed{\times 5}$—— 35 ——$\boxed{-7}$—— 28

Exercise 4   In the following problems

(a) write down the process

(b) write down the equation

(c) write down the inverse process

(d) use the inverse process to find unknown number in the equation.

1   8 times a number plus 5 equals 29. What is the number?

2   3 times a number plus 7 equals 13. What is the number?

3   9 times a number plus 4 equals 49. What is the number?

4   4 times a number take away 7 equals 25. What is the number?

5   5 times a number take away 2 equals 48. What is the number?

6   7 times a number take away 8 equals 34. What is the number?

7   Think of a number between 1 and 10. Multiply by 6. Add on 9.

Information ➪   Finding the unknown number in an equation is known as **solving the equation**.

Example   **Solving equations**
Solve $5x+3 = 48$.

Process: $x \longrightarrow \boxed{\times 5} \longrightarrow 5x \longrightarrow \boxed{+3} \longrightarrow 5x+3$

Inverse process: $5x+3 \longrightarrow \boxed{-3} \longrightarrow 5x \longrightarrow \boxed{\div 5} \longrightarrow x$

Equation: $5x+3 = 48$

Do $\boxed{-3}$ to each side: $\boxed{-3}\ \boxed{-3}$

$5x = 45$

Do $\boxed{\div 5}$ to each side: $\boxed{\div 5}\ \boxed{\div 5}$

$x = 9$

Exercise 5   Solve the following equations. Set out your working as shown in the example above.

1 $2x+4 = 10$	2 $3x+7 = 22$	3 $5x+2 = 7$
4 $4y+4 = 12$	5 $6a+3 = 27$	6 $3p+2 = 14$
7 $8x-2 = 6$	8 $3x-4 = 11$	9 $6x-2 = 10$
10 $3r-2 = 10$	11 $7t-4 = 17$	12 $10y-2 = 18$
13 $7x+5 = 26$	14 $9r-4 = 41$	15 $3c-8 = 19$
16 $9t+8 = 53$	17 $6d-19 = 11$	18 $7b+15 = 78$
19 $8c-7 = 65$	20 $7b+15 = 50$	21 $6f+17 = 71$

Information ⇨ Not all equations are as simple as the ones you have been solving. Sometimes there are letters as well as numbers on the right-hand side of the equation. To solve these you must apply the same rules as you have been using. You must rearrange the equation to have only letters on the left and only numbers on the right. You can do this using the inverse process.

Remember that **solving an equation** just means finding the unknown number which is represented by a letter in the equation. Each step should make the equation simpler and lead to the solution.

Examples   **Solving equations**

1   Solve $5x+4 = 3x+10$

$$5x+4 = 3x+10$$
$$\boxed{-4} \qquad \boxed{-4}$$
$$5x = 3x+6$$
$$\boxed{-3x} \qquad \boxed{-3x}$$
$$2x = 6$$
$$\boxed{\div 2} \qquad \boxed{\div 2}$$
$$x = 3$$

2   Solve $7y-3 = 4y+15$

$$7y\ \ 3 = 4y+15$$
$$\boxed{+3} \qquad \boxed{+3}$$
$$7y = 4y+18$$
$$\boxed{-4y} \qquad \boxed{-4y}$$
$$3y = 18$$
$$\boxed{\div 3} \qquad \boxed{\div 3}$$
$$y = 6$$

To check the answer substitute back into the equation.

$x=3$

so $5x+4 = 19$ and $3x+10 = 19$

$y=6$

so $7y-3 = 39$ and $4y+15 = 39$

Exercise 6   Solve these equations and check each answer by substituting back into the equation.

1   $5x+3 = 2x+9$ 2 $9x+5 = 4x+15$ 3 $4x+9 = 2x+13$

4   $8a+3 = 5a+21$ 5 $15c+4 = 9c+16$ 6 $13s+3 = 8s+33$

7   $5x-3 = 3x+7$ 8 $9x-5 = 4x+15$ 9 $8x-7 = 2x+11$

10   $13c-9 = 6c+33$ 11 $5r-12 = r+4$ 12 $14p-7 = 8p+41$

13   $8x+5 = 3x+25$ 14 $9x-2 = 5x+22$ 15 $9y+2 = 3y+20$

16   $8f-7 = 3f+38$ 17 $10s-13 = 3s+29$ 18 $9v+5 = v+29$

19   $8n+2 = 6n+14$ 20 $4x-7 = 2x+17$ 21 $12y-3 = 3y+15$

22   $6t+7 = 4t+13$ 23 $9c-5 = 5c+11$ 24 $6d-19 = d+31$

25   $8l-7 = 3l+28$ 26 $16c+3 = 9c+31$ 27 $19z-8 = 14z+37$

28   $40s+9 = 30s+39$ 29 $95v-7 = 83v+41$ 30 $64g+7 = 36g+30$

31   $6p+7 = 4p+13$ 32 $47x-19 = 12x+16$ 33 $19p+8 = p+44$

34   $8p+3 = 6p+4$ 35 $7y+4 = 3y+6$ 36 $5x+9 = 3x+10$

37   $4x-1 = 2x+4$ 38 $8x-5 = 4x+1$ 39 $9x-4 = 5x+10$

40   $6x-1 = 2x$ 41 $7x-4 = 5x$ 42 $8x-3 = 4x+2$

43   $10x-13 = 6x+5$ 44 $10x-3 = 2x+3$ 45 $13y\ \ 5 = 9y+5$

Using a formula

Information ⇨ In an earlier unit you looked at the use of formulae. You substituted and calculated the value of the **subject** of the formula. In the formula $A = lb$, A is the subject. The letter which you need to find may not always be the subject of the formula.

Examples

Substituting into the formula $V = u + at$

1 Find V if $u = 7$, $a = 5$ and $t = 9$.

Substitute: $V = 7 + 5 \times 9$

Multiply before adding: $V = 7 + 45$

$V = 52$

2 Find u if $V = 37$, $a = 9$ and $t = 3$.

Substitute: $37 = u + 9 \times 3$

Multiply before adding: $37 = u + 27$

Letters on the left: $u + 27 = 37$

Take away 27: $\boxed{-27}\ \ \boxed{-27}$

$u = 10$

3 Find a if $V = 90$, $u = 36$ and $t = 6$.

Substitute: $90 = 36 + 6a$

Letters on the left: $6a + 36 = 90$

Take away 36: $\boxed{-36}\ \ \boxed{-36}$

$6a = 54$

Divide by 6: $\boxed{\div 6}\ \ \boxed{\div 6}$

$a = 9$

4 Find t if $V = 45$, $u = 9$ and $a = 3$.

Substitute: $45 = 9 + 3t$

Letters on the left: $3t + 9 = 45$

Take away 9: $\boxed{-9}\ \ \boxed{-9}$

$3t = 36$

Divide by 3: $\boxed{\div 3}\ \ \boxed{\div 3}$

$t = 12$

Example

Using a formula

Ihab has decided to sow grass seed for a new lawn. He has prepared a rectangular area measuring 10 metres by 6 metres and has a packet of seed which will cover an area of 42 m². He decides to work in 6 metre strips going across the 10 metre length. How far down the length of the garden will he be able to sow the grass seed before he runs out?

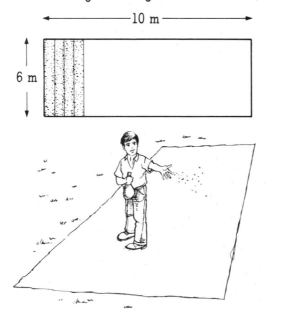

Ihab has enough seed to cover 42 m²
He is covering a rectangle 6 m broad. What will be the length of the rectangle?

$A = lb$
$A = 42$ m² and $b = 6$ m
You need to find l.

$$A = lb$$

Substitute: $42 = l \times 6$

$42 = 6l$

$6l = 42$

$\boxed{\div 6}\ \ \boxed{\div 6}$

$l = 7$

Ihab will be able to sow the seed for 7 metres along the prepared ground before he runs out.

Exercise 7

1   $V = IR$
    (a) Find V if $I=9$ and $R=25$.
    (b) Find I if $V=63$ and $R=7$.
    (c) Find R if $V=78$ and $I=6$.

2   $A = lb$
    (a) Find A if $l=12$ and $b=9$.
    (b) Find l if $A=96$ and $b=16$.
    (c) Find b if $A = 45$ and $l=9$.

3   $F = ma$
    (a) Find F if $m=9$ and $a=6$.
    (b) Find m if $F=80$ and $a=10$.
    (c) Find a if $F=90$ and $m=15$.

4   $P = VI$
    (a) Find P if $V=240$ and $I=3$.
    (b) Find V if $P=480$ and $I=4$.
    (c) Find I if $P=1200$ and $V=240$.

5   $E = mgh$
    (a) Find E if $m=4$, $g=10$ and $h=2$.
    (b) Find m if $E=200$, $g=10$ and $h=4$.
    (c) Find h if $E=240$, $g=10$ and $h=8$.

6   $v = u+at$
    (a) Find v if $u=9$, $a=8$ and $t=7$.
    (b) Find u if $v=39$, $a=7$ and $t=4$.
    (c) Find a if $v=54$, $u=18$ and $t=6$.
    (d) Find t if $V=47$, $u=19$ and $a=4$.

7   $l = a+(n-1)d$
    (a) Find l if $a=3$, $n=10$ and $d=5$.
    (b) Find a if $l=35$, $n=7$ and $d=4$.
    (c) Find d if $l=50$, $a=2$ and $n=9$.

8   $A = \pi rs$, $\pi=3.14$
    (a) Find A if $r=6$, $s=10$
    (b) Find r if $A=60$, $s=3$
    (c) Find s if $A=40$, $r=4$

9   $C = \dfrac{5\,(F-32)}{9}$
    (a) Find C if $F=50$
    (b) Find F if $C=25$
    (c) Check your answers against a thermometer

10 $T = \dfrac{12\,(D-d)}{L}$
    (a) Find T if $D=18$, $d=10$, and $L=4$
    (b) Find D if $T=6$, $L=14$, and $d=9$
    (c) Find d if $T=3$, $L=20$, and $D=6$

11 Ben is buying vinyl to lay on his kitchen floor which is rectangular in shape. He remembers the sales assistant telling him that a 3 metre roll of vinyl would fit exactly across his kitchen floor and that his bill would be for 15 m² of flooring. What length of vinyl is being delivered?

12 Robin is a sheep farmer. He has 1 large field which he keeps for fattening up his sheep before taking them to market. The field is rectangular in shape and measures 280 m long and 150 m wide. He has a portable electric fence 150 m long which he uses to divide the field. He needs an area of 3750 m² for 40 sheep. How far down the field must he put the electric fence?

13 An insurance company uses the formula $S = 6000+500y$ to calculate the salaries of their clerks, where S is the salary and y is the number of years the clerk has been working for the company.
    (a) What is George's salary if he has been with the company 2 years?
    (b) Shabanna earns £8500. How many years has she worked with the company?

Straight line graphs

Information ➥ The equations you have looked at so far have only had 1 **variable** (usually *x*). This means there is only 1 letter standing for an unknown number. You are now going to look at equations with 2 variables. They have 2 different letters standing for 2 different unknown numbers. There are many possible pairs of numbers which fit each equation. The solution to these equations can be shown in graphs.

Example

Graph of $x+y = 9$

George and Patrick are going to a party and have 9 cans of beer to share. Use a graph to show how they could be shared.

You can say George has *x* cans and Patrick has *y* cans. The possible solutions are shown in the table.

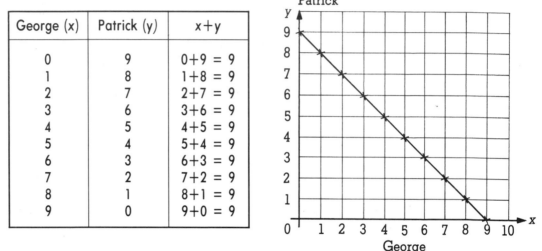

George (x)	Patrick (y)	x+y
0	9	0+9 = 9
1	8	1+8 = 9
2	7	2+7 = 9
3	6	3+6 = 9
4	5	4+5 = 9
5	4	5+4 = 9
6	3	6+3 = 9
7	2	7+2 = 9
8	1	8+1 = 9
9	0	9+0 = 9

The points on the graph show all the possible ways that the cans can be shared if George and Patrick each have only whole cans. It is also possible that they may share 1 can and each have a fraction of a can. On the graph fractions are found between the numbers on each axis. When fractions can be included in the solution you can join up the points with a straight line.

Example

Graph with whole numbers only

Julia's cat has had 5 kittens. Sajid and Helen both want the kittens. How could they be shared?

You can say Sajid has *x* kittens and Helen has *y* kittens. The possible solutions are shown in the table. You cannot join up the points in this graph because 1 kitten cannot be shared out!

Sajid (x)	Ihab (y)	x+y = 5
0	5	0+5 = 5
1	4	1+4 = 5
2	3	2+3 = 5
3	2	3+2 = 5
4	1	4+1 = 5
5	0	5+0 = 5

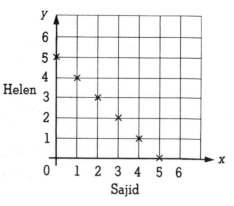

Exercise 8

1 Susan and Syeda have won 12 bottles of wine in a prize draw. Draw a table and graph to show how the prize could be shared. Remember to label the graph with the equation you have drawn.

2 Video tapes are cheaper when bought in packs of ten. Paul and Liam buy 10 video tapes to share. Draw a table and graph to show how they could share the tapes. Remember to label the graph with the equation you have drawn.

3 Mandy and Sarah help at the local hospital. A grateful patient has given them a bunch of 8 roses. Draw a table and graph to show how they could share the roses. Remember to label the graph with the equation you have drawn.

4 Brian and Mumtaz buy a bag containing 6 packets of crisps. Draw a table and graph to show how they could share out the crisps. Label the graph with the equation you have drawn.

5 Morag has 7 baby gerbils which she wants to give away to 2 of her friends. Draw a table and graph to show how they could share the gerbils. Label the graph with the equation you have drawn.

6 Roger has 11 bars of chocolate to share with his friend Fiaz. Draw a table and graph to show how the chocolate could be shared. Label the graph with the equation you have drawn.

Information ↪   In the last exercise you drew the graphs which showed the solutions to equations. In each case the points lay in a straight line. Only 2 points are needed to draw a straight line, but you need to plot at least 3 points to check that the points do lie in a straight line and not in a curve.

Examples

Graphical solution of an equation

1   Draw the graph which shows the solution of the equation $x+y = 9$.

Choose any 3 pairs of numbers which add up to 9.

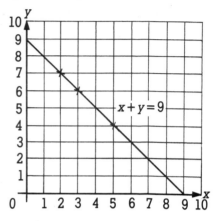

x	y	$x+y = 9$
3	6	$3+6 = 9$
5	4	$5+4 = 9$
2	7	$2+7 = 9$

Plot the points (3,6), (5,4) and (2,7).
Join them up in a straight line.
Extend the line.

2   Draw the graph which shows the solution of the equation $x-y = 2$.

Choose any 3 pairs of numbers which have a difference of 2. For example $7-5 = 2$.
Make x the larger number.

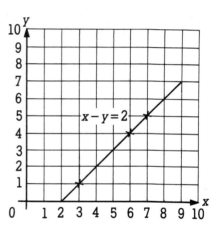

x	y	$x-y = 2$
7	5	$7-5 = 2$
3	1	$3-1 = 2$
6	4	$6-4 = 2$

Plot the points (7,5), (3,1) and (6,4).
Join them up in a straight line.
Extend the line.

3   Draw the graph which shows the solution of the equation $y = 4$.

Choose any 3 points which have 4 as the y coordinate. The y coordinate is the second number in a pair of coordinates.

x	y
1	4
3	4
5	4

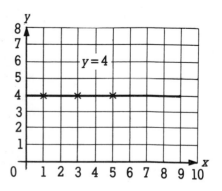

Plot the points (1,4), (3,4) and (4,4).
Join them up in a straight line.
Extend the line.

Exercise 9   Draw the graph of each of the equations below. Set out your working as in the examples on the previous page. Draw each graph on a separate diagram.

1 $x+y = 8$ 2 $x+y = 3$ 3 $x+y = 5$

4 $x-y = 3$ 5 $x-y = 1$ 6 $x-y = 4$

7 $y = 5$ 8 $y = 3$ 9 $y = 8$

10 $x = 3$ 11 $x = 6$ 12 $x = 4$

The straight line graphs given in questions **13-20** have not been labelled. Write down the equation of each graph.

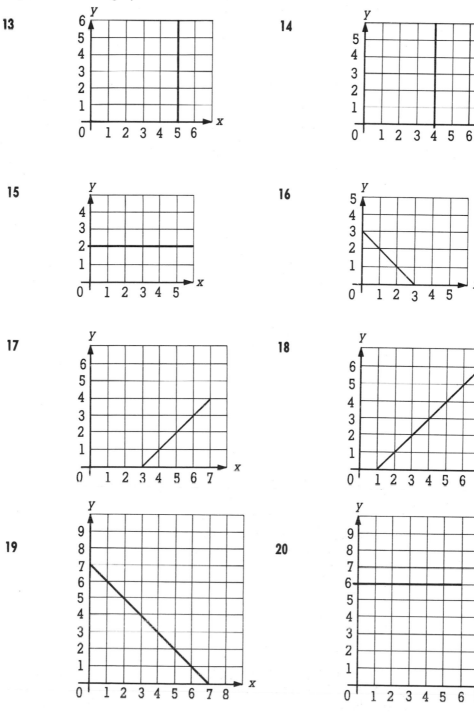

13

14

15

16

17

18

19

20

Information ➥ Equations with 2 variables can be written in different ways. Sometimes y is the subject of the equation and sometimes the x and y terms are combined on the same side. You have to use different methods to find pairs of numbers which give the solution to these equations.

Example

Equation with y as the subject
Show the solution of the equation $y = 2x - 1$ on a graph.

Since y is the subject of the equation you need to choose values for x and then substitute to find the values of y.

x	2x−1	y
1	2×1−1	1
2	2×2−1	3
3	2×3−1	5

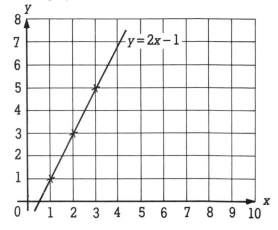

Plot the points (1,1), (2,3) and (3,5).
Join them up with a straight line.
Extend the line.

Example

Equation with x and y combined
Show the solution to the equation $2x + 3y = 12$ on a graph.

In this case you need to choose 3 values of one variable, x, and then substitute. You will then have to solve an equation to find the corresponding values of the other variable, y. The values of y will not always be whole numbers.

Substitute $x = 0$:	$2 \times 0 + 3y = 12$
Multiply before adding:	$3y = 12$
Divide by 3:	$y = 4$

Substitute $x = 2$:	$2 \times 2 + 3y = 12$
Multiply before adding:	$4 + 3y = 12$
Subtract 4:	$3y = 8$
Divide by 3:	$y = 2\frac{2}{3}$

Substitute $x = 3$:	$2 \times 3 + 3y = 12$
Multiply before adding:	$6 + 3y = 12$
Subtract 6:	$3y = 6$
Divide by 3:	$y = 2$

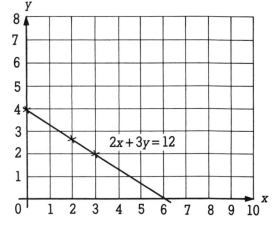

Plot the points (0,4), (2,2⅔) and (3,2).
Join the points up with a straight line.
Extend the line.

Exercise 10   Show the solution to each of the equations in questions **1-12** on separate graphs.

1 $y = 2x+1$ 2 $y = 3x+2$ 3 $y = x$

4 $y = 3x+1$ 5 $y = x+3$ 6 $y = 2x+3$

7 $y = 3x-2$ 8 $y = x-1$ 9 $y = 3x-1$

10 $y = 2x$ 11 $y = 2x-2$ 12 $y = 4x-3$

Plot the solution to each set of equations in questions **13-15** on the same graph.

13(a) $y = x$ (b) $y = x+1$ (c) $y = x+2$

14(a) $y = 2x$ (b) $y = 2x+1$ (c) $y = 2x+2$

15(a) $y = 3x$ (b) $y = 3x+1$ (c) $y = 3x+2$

16 Write down anything you notice about the sets of graphs in questions **13-15.**

Plot the solution to the equations in questions **17-22** on separate graphs.

17 $2x+y = 10$ 18 $3x+y = 12$ 19 $2x+y = 8$

20 $4x+2y = 14$ 21 $x+2y = 6$ 22 $3x+2y = 12$

Write down the equation of the lines shown in the graphs below. You may find it useful to draw a table and put in at least 3 points from the graph.

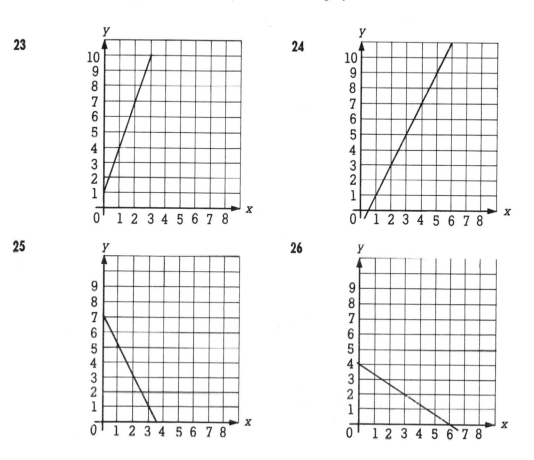

23

24

25

26

Simultaneous equations

Information ↩   When you draw a graph to show the solutions to an equation the line of the graph represents the pairs of numbers which are all solutions to this equation. If you draw 2 lines, each giving the solution to a different equation and the lines cross, the point where the lines cross is important. That point is the pair of numbers which gives the solution to both equations at the same time. The 2 equations are called **simultaneous equations**.

Example

Simultaneous equations

Find the pair of numbers which are the solution to the simultaneous equations $x+y = 6$ and $2x+y = 8$.

Draw the graphs of the solution to each of these equations on the same diagram.

The point where the 2 lines cross is the solution to the simultaneous equations. The solution is (2,4)

Check by substituting $x=2$ and $y=4$, into $x+y = 6$ and then into $2x+y = 8$.

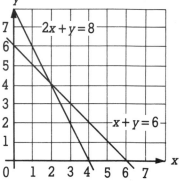

Exercise 11   Write down the solution to these simultaneous equations.

1

2

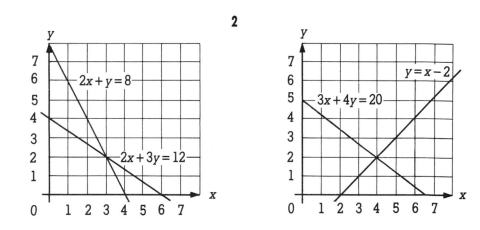

In questions **3-12** draw the lines which show the solutions to each pair of equations on the same graph. Write down the solution to each pair of simultaneous equations. Check your answer by substituting.

3 $x+y = 3$ and $2x+y = 4$ **4** $2x+3y = 9$ and $x+2y = 5$

5 $4x+y = 13$ and $2x+y = 9$ **6** $x+2y = 8$ and $2x+y = 10$

7 $x+y = 10$ and $x-y = 0$ **8** $2x+y = 14$ and $x-y = 4$

9 $4x-y = 4$ and $2x+y = 8$ **10** $5x+y = 10$ and $5x-y = 0$

11 $2x+y = 13$ and $5x-2y = 1$ **12** $4x+y = 8$ and $6x-y = 2$

Inequations

Information ➩ So far you have only considered equations. However **inequations** are also important. There are 4 symbols which are used in inequations. They are

$<$ **less than** $>$ **greater than**

$\leqslant$ **less than or equal to** $\geqslant$ **greater than or equal to**

Inequations are solved by applying the same rules which you have already used to solve equations.

Examples   **Inequations**

1 The speed limit in built-up areas is 30 mph. Write this as a word inequation.

$$\text{Speed} \leqslant 30 \text{ mph}$$

2 In the heavyweight class of boxing, boxers must weigh a minimum of 81 kg. Write this as a word inequation.

$$\text{Weight} \geqslant 81 \text{ kg}$$

3 Solve the inequations $3a+5 > 17$ and $5y-4 < 11$

$3a+5 >$	17		$5y-4 <$	11
$\boxed{-5}$	$\boxed{-5}$		$\boxed{+4}$	$\boxed{+4}$
$3a >$	12		$5y <$	15
$\boxed{\div 3}$	$\boxed{\div 3}$		$\boxed{\div 5}$	$\boxed{\div 5}$
$a >$	4		$y <$	3

Exercise 12   Write each of these statements as a word inequation.

1 The speed limit on motorways is 70 mph.

2 The minimum age that you can legally buy alcohol is 18.

3 The minimum age that you can legally buy tobacco is 16.

4 Weightlifters in the flyweight category must not weigh more than 52 kg.

Write out the following pairs of quantities as inequations using the correct symbol.

5 4 cm and 7 cm   **6** 18 m and 14 m   **7** £5 and £7

8 3 km and 5 km   **9** 11 mm and 1 cm   **10** £1 and 95p

11 960 g and 1 kg   **12** 103 cm and 1 m   **13** £6.95 and 700p

14 13 inches and 1 foot   **15** 1 metre and 2 feet

Solve the following inequations

16 $2x+5 > 13$ **17** $6v+7 < 13$ **18** $5z+3 > 33$

19 $8c-5 > 51$ **20** $6t-3 > 39$ **21** $9y-7 < 20$

At the circus

The problem can be solved using simultaneous equations.

Alison and Niven bought   4 adults and 4 child's tickets for £12
The next woman bought   2 adult and 4 child's tickets for £8

You can say the adult ticket cost £x and the child's ticket cost £y.

So Alison and Niven paid   £4x+£4y   The equation is 4x+4y=12
The next woman paid   £2x+£4y   The equation is 2x+4y=8

To find 3 points satisfying 4x+4y = 12:

Substitute $x=0$
$4\times0 + 4y = 12$
$4y = 12$
$y = 3$

Substitute $x=1$
$4\times1 + 4y = 12$
$4 + 4y = 12$
$4y = 8$
$y = 2$

Substitute $x=2$
$4\times2 + 4y = 12$
$8 + 4y = 12$
$4y = 4$
$y = 1$

Point (0,3)   Point (1,2)   Point (2,1)

To find 3 points satisfying 2x+4y = 8:

Substitute $x=0$
$2\times0 +4y = 8$
$4y = 8$
$y = 2$

Substitute $x=1$
$2\times1 + 4y = 8$
$2 + 4y = 8$
$4y = 6$
$y = 1\frac{1}{2}$

Substitute $x=2$
$2\times2 + 4y = 8$
$4 + 4y = 8$
$4y = 4$
$y = 1$

Point (0,2)   Point (1,1½)   Point (2,1)

On the same diagram plot and join up the 3 points for each line.

The lines cross at the point (2,1).

x is 2 so an adult ticket costs £2.
y is 1 so a child's ticket costs £1.

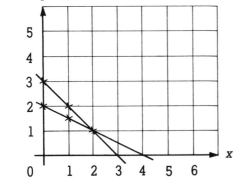

Check by going back to the problem and substituting the values of £2 for an adult ticket and £1 for a child's ticket.

Alison and Niven: 4 adult plus 4 child's gives 4×£2 + 4×£1 = £12
The next woman: 2 adult plus 4 child's gives 2×£2 + 4×£1 = £8

Alison and Niven did pay £12 and the next woman did pay £8 so the values are correct.

How much does Pauline owe Alison?

Using equations

Exercise 13

1 The Savra family bought 3 adult and 2 child's tickets for £18.

The Williams family bought 4 adult and 3 child's tickets for £25.

What is the cost of the adult ticket? What is the cost of the child's ticket?

2 Tony, the Baker, sells wholemeal and white loaves. 3 wholemeal and 2 white loaves cost £2.19. 2 wholemeal and 3 white loaves cost £2.16. How much does each type cost?

6 Pythagoras

In this unit it is assumed that you know how to:

1 square numbers

2 estimate the square root of numbers

3 find the square root of numbers using a calculator

Exercise 0 will help you check. Can you answer all the questions?

Exercise 0

Squaring numbers

1 (a) Copy and complete this table

x	1	2	3	4	5	6	7	8	9	10	11	12	13	14	15	16	17	18	19	20
x^2	1	4	9	16	25			64			121				225					400

(b) Using suitable scales draw a graph of x against x^2.

Estimating square roots

Use your graph to estimate the following square roots correct to 1 decimal place.

2 $\sqrt{90}$ 3 $\sqrt{40}$ 4 $\sqrt{60}$ 5 $\sqrt{130}$

6 $\sqrt{45}$ 7 $\sqrt{75}$ 8 $\sqrt{175}$ 9 $\sqrt{280}$

10 $\sqrt{360}$ 11 $\sqrt{293}$ 12 $\sqrt{187}$ 13 $\sqrt{43}$

Without using your graph, estimate the square root of the following numbers.

14 $\sqrt{77}$ 15 $\sqrt{190}$ 16 $\sqrt{375}$ 17 $\sqrt{54}$

18 $\sqrt{222}$ 19 $\sqrt{162}$ 20 $\sqrt{305}$ 21 $\sqrt{92}$

Using a calculator to find square roots

22 Use a calculator to check the accuracy of your answers.

Investigation

Happy numbers

23 is a happy number. Follow these steps:

Square its digits and then add the squares $2^2+3^2 = 4+9 = 13$
Repeat this process until you get a single digit answer $1^2+3^2 = 1+9 = 10$
                                                    $1^2+0^2 = 1+0 = 1$

Happy numbers all finish with 1.
Sad numbers do not finish with 1.

42 is a sad number
$4^2+2^2=16+4=20$
$2^2+0^2= 4+0=4$

Do you think there are more happy numbers or sad numbers?

Crisis

Commons fury over Tripoli raid

There were furious exchanges in the House of Commons last night as MPs discussed the use of British-based F111 planes in the US attack on Tripoli. The Prime Minister was severely criticised by Neil Kinnock, leader of the opposition, for allowing the F111s to be used when France and Spain both distanced themselves from the attack by refusing permission to fly in their airspace.

In April 1986 a major international crisis developed when American planes based in Britain attacked Libya. There were also heated debates in the House of Commons as to whether the British Government should have allowed the planes to be used. Britain's position was in contrast to France and Spain who refused to allow the planes to fly in their airspace.

The route taken by the American planes was approximately as shown.

The direct distance from their base in Britain to Tripoli, in Libya is 2500 km. The planes flew 1400 km along the first part of the journey to avoid France and Spain then turned through 90 degrees to head for Tripoli. What is the extra distance they have to fly because of the decision by France and Spain?

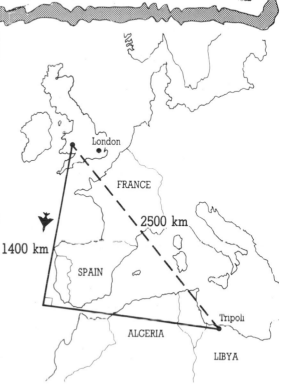

If the planes left Britain with enough fuel to fly 7500 km would this be enough for them to complete their mission? (Remember they had to return to their base in Britain by the same route.)

This unit will help you to solve problems like this one.

Discovering a formula

Investigation   (a) Draw each of these 6 right-angled triangles accurately on 1 centimetre squared paper. Label the sides as shown. You are going to find a formula for the length of the longest side which is labelled c.

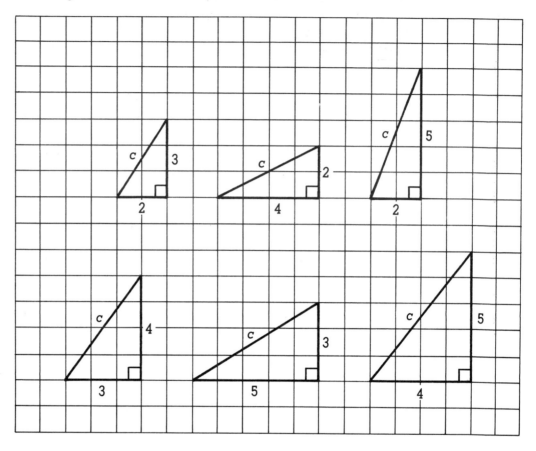

(b) Copy and complete the table below. Use a ruler to measure side c, correct to the nearest millimetre, for each of the 6 triangles. Find c^2 correct to the nearest whole number. The first line has been completed as a guide.

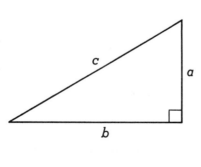

a	b	c	a^2	b^2	c^2	$a^2 + b^2$
2	3	3.6	4	9	13	13
4	2					
2	5					
3	4					
5	3					
4	5					

(c) Draw 2 more right-angled triangles, choosing lengths for a and b (the 2 shorter sides) not already used in the table. Put your values in the last 2 lines of the table.

(d) What conclusions can you draw from the table?

Estimating sides

Information ↩ Pythagoras was a Greek mathematician who first discovered the relationship between the sides of any right-angled triangle.

Side c has a special name. It is called the **hypotenuse**. It is the **longest side** in a right-angled triangle and is always **opposite the right-angle**.

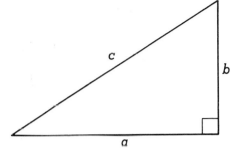

In any triangle the sum of 2 sides is always greater than the third side. We can use this to estimate the length of sides in a right-angled triangle.

Examples **Estimating the length of sides**
Estimate the length of the side a, b or c in each of these triangles.

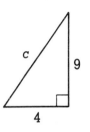

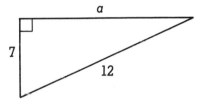

c is the hypotenuse
c must be greater than 9
c must be less than $4+9 = 13$
c is between 9 and 13

Estimate 11

a is shorter than the hypotenuse
a must be less than 12
$a+7$ must be greater than 12
a must be greater than 5
so a is between 5 and 12
Estimate 9

Exercise 1 Estimate the length of side a, b or c in each of these right-angled triangles. Set out your working as shown in the examples above.

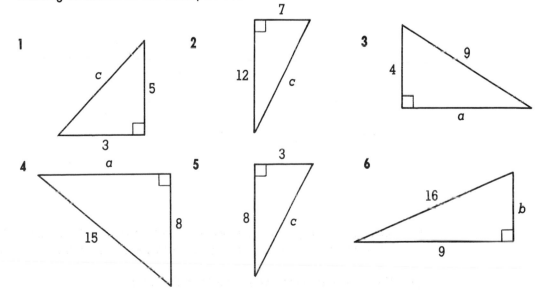

Working out the hypotenuse

Information ⇨ Pythagoras theorem is used to find the length of 1 side of a right-angled triangle when the other 2 sides are known. You may have worked out this theorem from the investigation at the start of this unit. The theorem states that:

The length of the hypotenuse in a right-angled triangle is found by using the formula

$c^2 = a^2 + b^2$.

where c is the length of the hypotenuse.

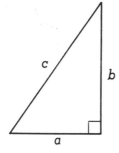

Example **Finding the length of the hypotenuse**
Find the length c in the right-angled triangle below.

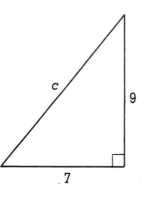

Estimating:
c lies between 9 and 16 (9+7)
$9 < c < 16$

$c^2 = 9^2 + 7^2$

Using the formula:
$c^2 = 81 + 49$
$c^2 = 130$
$c = \sqrt{130}$ (lies between 11 and 12)
$c = 11.4$

Exercise 2 Find the length of the hypotenuse of each of these right-angled triangles. Set out your working like the example above.

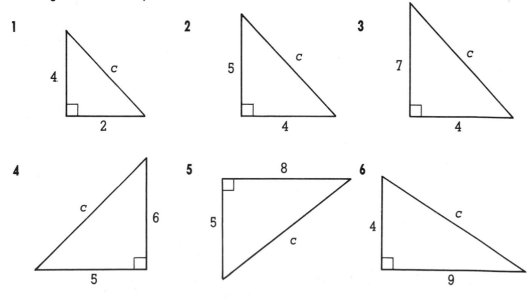

Working out the shorter sides

Information ⇔ To find the length of one of the shorter
sides in a right-angled triangle, use the
same formula but rearranged

$$b^2 = c^2 - a^2$$
$$a^2 = c^2 - b^2$$

where c is the length of the hypotenuse,
a and b are the two shorter sides.

If you know the length of the hypotenuse
and the length of one of the shorter sides
you can calculate the length of the other
side.

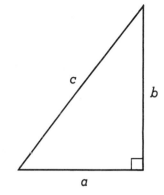

Example

Finding the length of a shorter side
Find the length of a in the right-angled triangle below.

Estimating:
the hypotenuse $c = 17$
a lies between 8 ($= 17 - 9$) and 17
$8 < a < 17$

Using the formula:
$a^2 = 17^2 - 9^2$
$a^2 = 289 - 81$
$a^2 = 208$
$a = \sqrt{208}$ (lies between 14 and 15)
$a = 14.4$

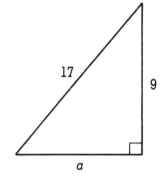

Exercise 3   Find the missing length in each of these right-angled triangles. Set out your working like
the example above.

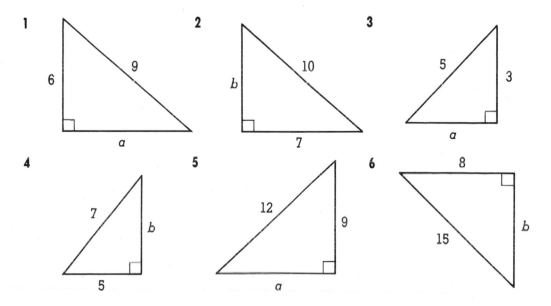

Mixed examples

Information ⇨

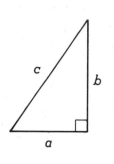

c is the length of the hypotenuse
a and b are the lengths of the shorter sides.
To find the **hypotenuse** use $c^2 = a^2 + b^2$
To find a **shorter side** use $a^2 = c^2 - b^2$
$$b^2 = c^2 - a^2$$

Examples

Pythagoras theorem
Find the length of the missing side in these right-angled triangles.

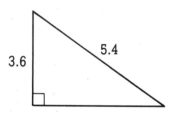

The missing side is the hypotenuse, so call this c.
Estimating:
c lies between 6.5 and 11.2 (=6.5+4.7)
6.5<c<11.2

Using the formula:
$c^2 = 4.7^2 + 6.5^2$
$c^2 = 22.09 + 42.25$
$c^2 = 66.34$
$c = \sqrt{66.34}$ (between 8 and 9)
$c = 8.02$

The missing side is a shorter side, so call this a.
Estimating:
a lies between 1.8 (=5.4−3.6) and 5.4
1.8<a<5.4

Using the formula:
$a^2 = 5.4^2 - 3.6^2$
$a^2 = 29.16 - 12.96$
$a^2 = 16.2$
$a = \sqrt{16.2}$ (between 4 and 5)
$a = 4.02$

Exercise 4

Find the missing length in each of these right-angled triangles. Always make an estimate first of all.

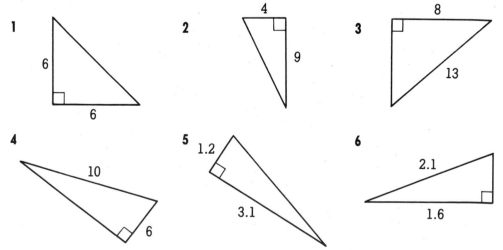

7

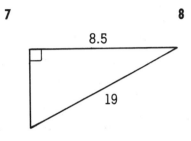

8.5

19

8

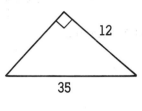

14.2

17

9

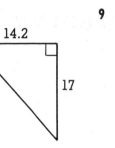

12

35

10

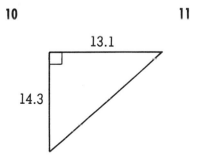

13.1

14.3

11

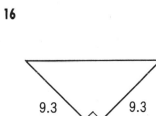

12

13

12

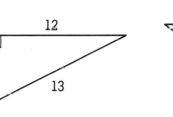

7.1

6.2

13

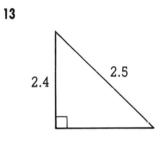

2.4

2.5

14

9.6

13.4

15

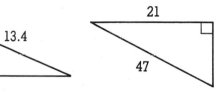

21

47

16

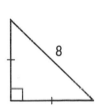

9.3   9.3

17

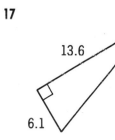

13.6

6.1

18

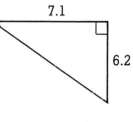

42

67

19

8

20

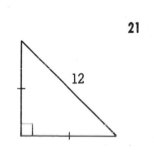

12

21

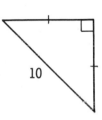

10

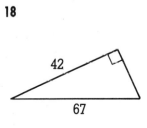

Pythagoras problems

Information ⇨   The following problems can all be solved by applying Pythagoras theorem. Here is a checklist to help you solve them.

1   Make a sketch of the problem.

2   Mark all right-angles and lengths.

3   Make a separate sketch of the right-angled triangle which contains the missing side you want to find.

4   Make an estimate of the missing side.

5   Apply Pythagoras theorem.

6   Think .... *Is my answer sensible? How does it compare with my estimate?*

Example   **Problem solving with Pythagoras theorem**
The Scottish flag has a white diagonal cross on a blue background. A flag measures 40 cm by 20 cm. What length of material is needed for the cross? (Assume that the cross is made of two diagonal lengths of material which overlap in the centre.)

Draw a sketch:

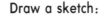

Draw a right-angled
triangle with the
information marked:

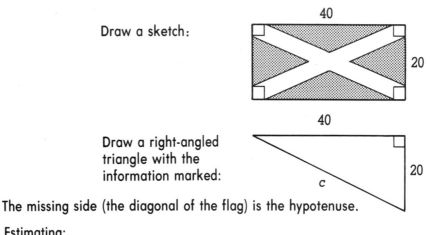

The missing side (the diagonal of the flag) is the hypotenuse.

Estimating:
c lies between 40 and 60 $40 < c < 60$

Using the formula:
$c^2 = 40^2 + 20^2$
$c^2 = 1600 + 400$
$c^2 = 2000$
$c = \sqrt{2000}$
$c = 44.7$ cm. This agrees with the estimate.

Length of material needed for the cross 2×44.7 cm $= 89.4$ cm

Exercise 5

1   A rectangle measures 15 cm by 11 cm.
    What is the length of the diagonal?

2   The sketch shows an isosceles triangle
    8 cm high with a base of 6 cm. What is
    the length of each side?

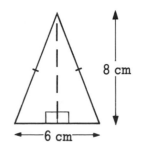

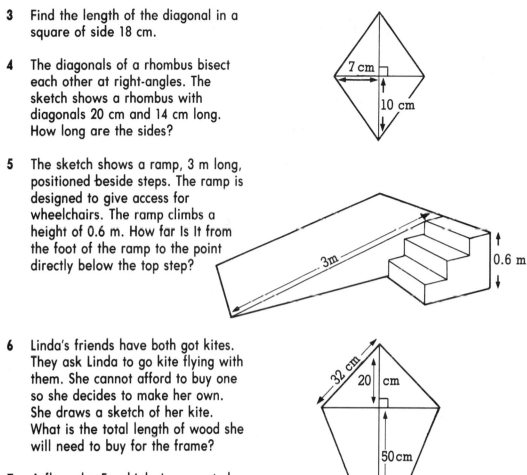

3 Find the length of the diagonal in a square of side 18 cm.

4 The diagonals of a rhombus bisect each other at right-angles. The sketch shows a rhombus with diagonals 20 cm and 14 cm long. How long are the sides?

5 The sketch shows a ramp, 3 m long, positioned beside steps. The ramp is designed to give access for wheelchairs. The ramp climbs a height of 0.6 m. How far Is It from the foot of the ramp to the point directly below the top step?

6 Linda's friends have both got kites. They ask Linda to go kite flying with them. She cannot afford to buy one so she decides to make her own. She draws a sketch of her kite. What is the total length of wood she will need to buy for the frame?

7 A flagpole, 5 m high, is supported by wire guys, each 5.6 m long, attached to the top of the pole. How far from the foot of the pole are the guys anchored?

8 The sketch shows a section of ornamental fencing which is sold in 2 m lengths. The zig-zag section is made from one long piece of metal.

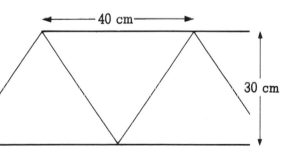

(a) What is the length of each diagonal?
(b) What length of metal is needed to make the complete zig-zag section for each 2 m length of fence?

9 The instructions attached to a ladder 4 m long are as follows:
The foot of the ladder must be placed between 1.5 m and 2 m from the foot of the wall.
What is the maximum height that the ladder will reach?

10 Tricia decides to try and save money by doing some rewiring herself instead of paying an electrician. She has to connect 2 electrical sockets at opposite corners of a room measuring 6 m by 4.5 m. She can run a cable either diagonally under the floor or round 2 sides of the room behind the skirting board. How much less cable will she use if she runs the cable diagonally under the floor?

11 The diagonal of a square is 25 cm long. How long are the sides?

12 A ship leaves port and sails on the following course. 35 km East then 40 km North then 10 km West. The captain then returns straight to port. How far, to the nearest kilometre, has ths ship sailed?

13 The diagram shows one of a series of wooden supports for a roof. The width of the building is 12 m, the height of the roof is 3 m and the sloping roof is 7 m long. What is the length by which the roof overhangs the building?

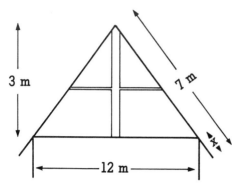

14 Peter is trying to design a ramp to fit exactly on top of a flight of 10 stairs, so that his mother can go up the steps in her wheelchair.

A section of the stairs is shown in the diagram with the ramp in place. How long must Peter make the ramp?

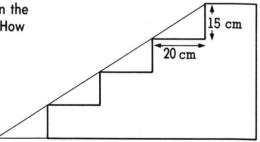

15 The sketch shows a simple bridge designed to span a 4 m gap. The diagonal supports meet in the middle of the bridge and are fixed to the sides of the gap 1.5 m below the surface. How long is each diagonal?

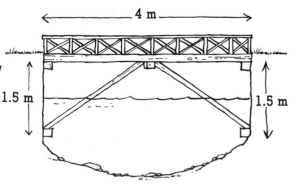

Circle and Pythagoras

Information ⇨ Right-angled triangles crop up in a variety of places. You can even find them in connection with the circle. Here are 3 facts connecting right-angled triangles and the circle. In each case Pythagoras theorem can be used to solve the problems arising.

1 The **angle in a semi-circle** is a right-angle.

2 A **radius meets any tangent** to the circle at a right-angle.
   A tangent is a straight line which **touches** the circle at **1 point only.**

3 A **radius meets a chord** at a right-angle if the **chord is bisected.**
   To bisect means to cut in **half.**

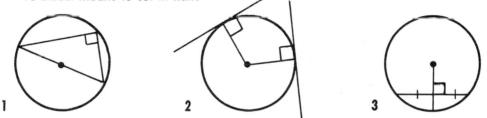

Try to prove the 3 facts yourself. For each fact accurately draw at least 6 different cases to see if the statement is true each time. For example, for fact 1 you would draw 6 triangles inside 6 semi-circles and measure the angle at the circumference.

Example   **Angle in a semi-circle**
Calculate the length of the diameter in this circle.

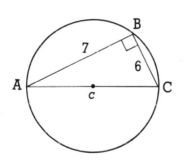

Angle ABC is a right-angle (Fact 1).
The diameter is the hypotenuse of triangle ABC.
Use Pythagoras theorem to find the hypotenuse.

Estimating:
$7 < c < 13$

Using the formula:
$c^2 = a^2 + b^2$
$c^2 = 6^2 + 7^2$
$c^2 = 36 + 49$
$c^2 = 85$
$c = \sqrt{85}$
$c = 9.2$ cm

Example   **Radius and Tangent**
Calculate the length of the tangent QR in the diagram below

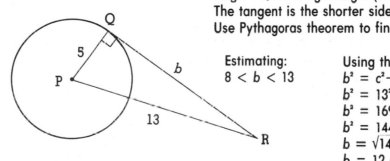

Angle PQR is a right-angle (Fact 2).
The tangent is the shorter side b, of triangle PQR.
Use Pythagoras theorem to find the shorter side.

Estimating:
$8 < b < 13$

Using the formula:
$b^2 = c^2 - a^2$
$b^2 = 13^2 - 5^2$
$b^2 = 169 - 25$
$b^2 = 144$
$b = \sqrt{144}$
$b = 12$ cm

Example

Radius and chord
Calculate the length of the chord in this circle.

Angle STU is a right-angle (Fact 3)
The length of ½ the chord is the shorter side, a, of triangle STU.
Use Pythagoras theorem to find the shorter side.

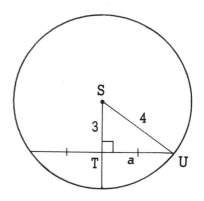

Estimating:
$1 < a < 4$

Using the formula:
$a^2 = c^2 - b^2$
$a^2 = 4^2 - 3^2$
$a^2 = 16 - 9$
$a^2 = 7$
$a = \sqrt{7}$
$a = 2.6$ cm

Length of chord is $2 \times 2.6 = 5.2$ cm

Exercise 6

Solve the problems below by applying the 3 facts and using Pythagoras theorem.

Sketch each of these diagrams. Below them write down which fact you are using and calculate the missing side.

1

2

3

4

5

6

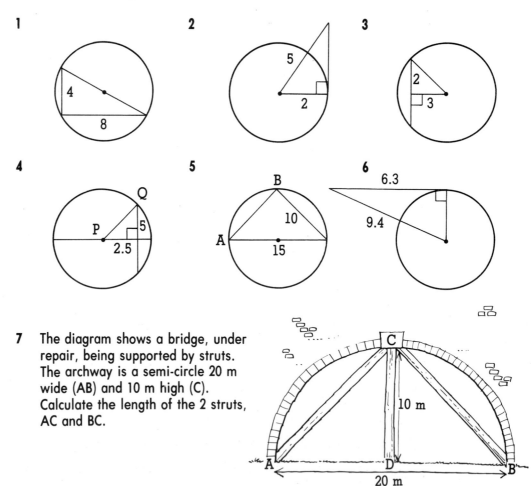

7 The diagram shows a bridge, under repair, being supported by struts. The archway is a semi-circle 20 m wide (AB) and 10 m high (C). Calculate the length of the 2 struts, AC and BC.

Crisis

Information ➩ Remember the route that the US planes took?
Here it is again.

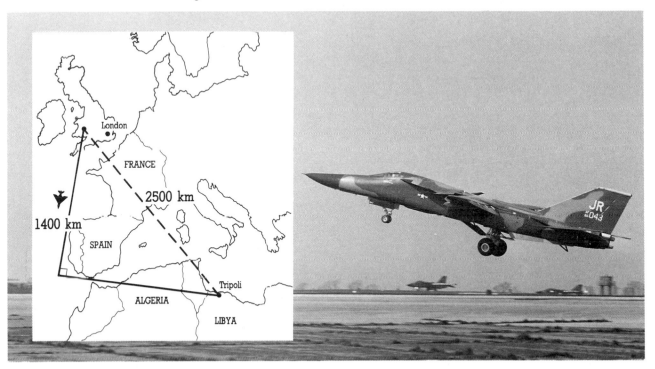

You can see that the route they took is the two short sides of a right-angled triangle.
The direct path is the hypotenuse.
The unknown part can be calculated.

Label the unknown side a.

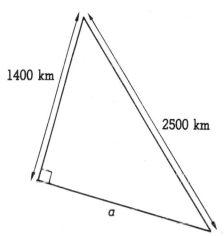

Estimating:
$1100 < a < 2500$

Using the formula:
$a^2 = 2500^2 - 1400^2$
$a^2 = 6\ 250\ 000 - 1\ 960\ 000$
$a^2 = 4\ 290\ 000$
$a = \sqrt{429\ 000}$
$a = 2\ 071.23$ km

The second leg of the route is approximately 2071 km.

Total length of route is $1400 + 2071 = 3471$ km.

So the extra distance is $3471 - 2500 = 971$ km.

Did they have enough fuel?

Total flying distance is $3471 \times 2 = 6942$ km

They have enough fuel for 7500 km, so they do have enough. Margin for error is
$7500 - 6942 = 558$ km.

Pythagorean triples

Information ⇨ In some of the exercises in this unit all the sides of the right-angled triangles came out as **whole numbers**. These are called **Pythagorean triples**.

Example **Basic triples**
Here are two basic Pythagorean triples. They are called **primitives**.

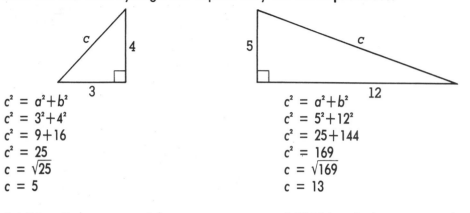

$c^2 = a^2 + b^2$
$c^2 = 3^2 + 4^2$
$c^2 = 9 + 16$
$c^2 = 25$
$c = \sqrt{25}$
$c = 5$

$c^2 = a^2 + b^2$
$c^2 = 5^2 + 12^2$
$c^2 = 25 + 144$
$c^2 = 169$
$c = \sqrt{169}$
$c = 13$

3,4,5 is a Pythagorean triple.   5,12,13 is a Pythagorean triple.

Any multiple of these would give another Pythagorean triple. These are called **multiples**.

6,8,10 would be a multiple since it is the Pythagorean triple 3,4,5 scaled up by a factor of 2.

Exercise 7 For each triangle in this exercise
(**a**) Calculate the length of the missing side.
(**b**) Write down the primitive of which it is a multiple.
(**c**) Write down the scale factor.

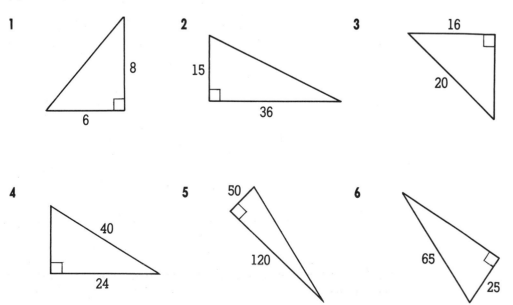

Investigation   Look at this right-angled triangle. You can calculate as many Pythagorean triples as you like by using 3 formulae, one formula for each side.

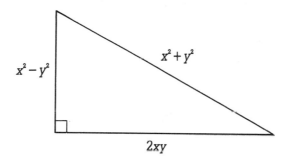

x and y are whole numbers and x is greater than y. Why must x be greater than y?

Examples   **Calculating a Pythagorean triple using the triangle above**

1   With $x = 2$ and $y = 1$

$$x^2 - y^2 = 2^2 - 1^2 = 4 - 1 = 3$$
$$2xy = 2 \times 2 \times 1 = 4$$
$$x^2 + y^2 = 2^2 + 1^2 = 4 + 1 = 5$$

So 3, 4, 5 is a Pythagorean triple

2   With $x = 3$ and $y = 1$

$$x^2 - y^2 = 3^2 - 1^2 = 9 - 1 = 8$$
$$2xy = 2 \times 3 \times 1 = 6$$
$$x^2 + y^2 = 3^2 + 1^2 = 9 + 1 = 10$$

So 6, 8, 10 is a Pythagorean triple.

Remember:
Pythagorean triples like 3,4,5 and 5,12,13 are called **primitive**.
Pythagorean triples like 6,8,10 and 10,24,26 are called **multiple**.

Copy and continue this table of Pythagorean triples. Do at least 8 more triples. How many primitives can you find?

x	y	$x^2 - y^2$	$2xy$	$x^2 + y^2$	**Primitive/Multiple**
2	1	3	4	5	Primitive
3	1	8	6	10	Multiple
3	2				
4	1				
4	2				
4	3				
5	1				
5	2				
5	3				
5	4				

Study the table and see if you can discover what must be true for x and y for the triple to be primitive.

Test your result with more values.

Write a report describing your conclusions and giving details of your tests.

---

7 Saving, spending and surviving

Do you remember? In this unit it is assumed that you know how to:

1 find a percentage of a sum of money
2 express one number as a percentage of another
3 use negative numbers in working out problems.

Exercise 0 will help you check. Can you answer all the questions?

Exercise 0

Percentages of sums of money

1 Find 8% of £200
2 Find 5% of £340
3 Find 4% of £525
4 Find 9% of £485
5 Find 15% of £1240
6 Find 7% of £6340
7 Find 4½% of £420
8 Find 7½% of £2640

Expressing one number as a percentage of another

9 Express 4 as a percentage of 20.
10 Express 6 as a percentage of 60.
11 Express 15 as a percentage of 60.
12 Express 35 as a percentage of 280.
13 Express 10.6 as a percentage of 106.
14 Express 19 as a percentage of 350 correct to 1 decimal place.
15 Express 38 as a percentage of 54 correct to 1 decimal place.
16 Express 93 as a percentage of 472 correct to 1 decimal place.

Using negative numbers to work out problems

17 Assad owes Fred £10 and Mary £16. He gets paid £35 a day for a Saturday job. How much is left after he pays his debts?

18 Harry has £60 in the bank. By how much is he overdrawn if he writes a cheque for £100?

19 Rose is overdrawn at the bank by £30. How much will she be in credit if she deposits £75?

Buying a car

Anna works for a company situated in a new industrial estate 15 miles away from the town in which she lives. Travelling to and from work by bus takes ages. She has just passed her driving test and is looking for a second hand car to buy for travelling to work in. After look around for several weeks she sees a Ford Fiesta for sale at £2000, and decides to buy it.

She has £500 saved up and can afford £60 per month to pay for the car.
It is for sale in a garage which offers HP terms of 5% interest plus 25% deposit.

How should she pay for the car? Here are some possibilities:

1   Save up until she has all the money.

2   Get a bank loan to make up the difference.

3   Take out a hire purchase agreement through the garage.

4   Apply for a credit card and use it to make up the difference.

If Anna decides to buy the car she will have further new expenses to consider:
  Road tax
  Insurance
  Petrol
  Repairs

This unit will help you make a sensible decision about problems like Anna's

Bank accounts

Information ➭ You will almost certainly use a bank at some time in your life. You may get your wages paid directly to the bank and have an account from which you take money as you need it or you may use a bank account to save money. There are 2 main types of bank account. These are called a **current account** and a **deposit account**.

Banks provide other services as well:

They buy and sell foreign currency.
They arrange mortgages.
They advise on investments.
You can also borrow money from a
bank, for example to buy a car.

Current accounts enable you to write **cheques** to pay bills, buy shopping, pay for plane tickets etc. This avoids the problem of carrying around large sums of money to pay bills.

Every month you get a **statement** from the bank which shows the **deposits** and **withdrawals** you have made and gives your **balance**.

Money paid into your account is known as a **deposit**. Money taken from your account is known as a **withdrawal** and the amount of money left in your account is called the **balance**.

Every time you write a cheque you should complete the **stub**. This is your record of the cheque and from which you can check your balance.

Example   **Completing a cheque and stub**
Martin Wesley had a balance of £231.56 before writing a cheque to pay his gas bill of £43.67. Here is the cheque to North Thames Gas and the stub which Martin completed. Can you see how he gets his new balance? Work out Martin's new balance.

DATE 13/12/87 TO North Thames Gas FOR Bill BALANCE BT. FWD. £231·56 DEPOSITS — TOTAL £231·56 THIS CHEQUE £ 43·67 BALANCE CD. FWD £187·89	**ANYBANK LTD** 80-02-73 DATE ___13 Dec 1987___ PAY North Thames Gas Forty three pounds and £43 — 67 67p only 0027.... 0002 M. Wesley.

Martin's new balance is £187.89 which should be written on the next stub.

Exercise 1

For questions **1-9** you will need a book of cheques similar to the one on the last page. You may have to draw them yourself or your teacher may have some. Write a cheque to pay for each of the following bills or purchases. Complete the stub on each cheque you write. Make sure to work out the new balance after each transaction. You start with £187.89 in your account on 15th June.

	Date	To	Amount
	15 June	Balance	£187.89
1	15 June	Electricity Board	£32.54
2	15 June	Marks and Spencer PLC	£16.72
3	18 June	Daley's Garage	£76.34
4	24 June	Gas Board	£43.25
	27 June	**Salary paid in to account**	**£575.82**
5	2 July	Star of Bengal Restaurant	£12.81
6	6 July	Britania Building Society	£175.86
7	10 July	Major Accident Insurance Ltd.	£6.38
8	12 July	Hi Fashion Boutique	£23.75
9	14 July	Cash	£75.00

10 Gino has a balance of £123.56 in his current account. He writes cheques to Harrods for £32.50, to C & As for £19.99 and to Tescos for £12.43. What is his new balance?

11 Betty has a balance in her current account of £84.50. She writes cheques to Arthur's Garage for £58.60 and Marks and Spencer for £32. What has happened to Betty's account?

12 John had a balance of £312.73 in his current account. He wrote 3 cheques, one for £56.10, one for £23.18 but he forgot to make a note of the third one. A week later he checked his balance and found that he had £194.92. What was the value of the third cheque?

13 Sheena has a balance of £76.82 in her current account. She writes three cheques for £23.76, £48.83 and £28.45. By how much is her account overdrawn?

Information ⮌ **Deposit accounts** have the advantage that **interest** is paid on the money you have in your account. This means that you get extra money depending on the amount that you have in your account.

Deposit accounts are more suitable for saving than for day to day use. Cheque books cannot be used with deposit accounts and you have to visit the bank to withdraw money. Each time money is deposited or withdrawn the bank teller records the transaction in your bank book.

Regular bills can be paid by monthly **standing order** from either a current or deposit account. This means that the bank will automatically pay the same amount, to the same place, from your account every month. A small charge is made for this service. Standing orders can be made for other intervals besides monthly.

Example   **Completing a bank book**
Here is a page from Anna's bank book.

Date		Initials	Deposited		Withdrawn		Balance	
15.9.87	B/Fwd	RC					246	75
15.9.87	Withdrawn	RC			25	00	221	75
19.9.87	Building Society (SO)	TR			74	36	147	39
23.9.87	Deposit	TR	50	00			197	39
25.9.87	Salary	FD	475	50			672	89
30.9.87								

B/Fwd means brought forward from the previous page. SO stands for standing order. If money is paid into her account it is recorded in the deposited column. If money is taken out of her account it is recorded in the withdrawn column. After each transaction the balance is calculated and recorded.

Exercise 2   Questions **1-5** give a list of Anna's bank transactions. Copy the bank book page shown above and then write in each of these transactions. Make sure to work out the balance after each one. You can make up the initials of the bank teller in each case.

1   30 September : Withdraw £75

2   2 October : Major Accident Insurance £12.65 (Standing Order)

3   7 October : Deposit £27.38

4   13 October : Transfer to current account £75

5   16 October : Interest paid £6.28

Simple interest

Information ⮐ **Interest** is the name given to the money your savings earn in a deposit account. The **interest rate** is the percentage per year which is used to calculate interest.

Example   **Simple interest for whole years**
When she was 21, Sarah was given £400 as a present from her parents. She put it in a deposit account at ·8% interest. How much interest will she have earned in 1 year?

Interest for 1 year 8% of £400 = £32

Each birthday she withdraws the interest which she has earned. How much interest will she have withdrawn in total after 3 years?

Interest for 3 years 3×£32 = £96

Example   **Simple interest for periods less than 1 year**
Calculate the interest on £160 invested at 15% per year over 4 months.

Interest for 1 year 15% of £160 = £24

Interest for 4 months $\dfrac{4}{12}$×£24 = £8

Exercise 3   Copy and complete the table below.

	Amount (£)	Interest rate	Time	Interest
	120	15%	2 years	£36
1	450	10%	1 years	
2	250	12%	5 years	
3	50	10%	6 months	
4	160	7%	9 months	
5	300	5%	3 months	
6	560	12½%	7 months	
7	65	2%	2½ years	
8	150	23.7%	18 months	
9	260	25.8%	9 months	

10   George invested £250 at 6% pa. How much interest did he get after 8 months?

11   Frank invested £960 in a building society for 7 months at 8½% pa. How much interest did he get?

12   Kirsten had £480 in a savings account for 6 months. For the first 3 months the interest rate was 4½% pa and then it increased to 5% pa. How much interest did she get after 6 months?

13   Jessie has £800 in a savings account. After 1 year her interest is £56. What is the interest rate?

14   Derek invested £560 for 6 months and got £16.80 interest. What is the interest rate per annum?

Compound interest

Information ➩ Most banks and building societies calculate **compound interest** and not simple interest. This means that interest is added onto the investment, say after one year, which then increases the amount on which interest is calculated for the next year and so on. The amount of money in an account is called **capital**. Building societies and banks often change their **interest rates**.

Example

Compound interest added once a year
If Sarah had left the £400 she was given in the bank and allowed the compound interest to accumulate over 3 years how much total interest would she have earned?

Interest is calculated on a whole number of pounds only.

Year 1	Capital	£400.00	
	Interest	£ 32.00	8% of £400 = £32
	Total	£432.00	

Year 2	Capital	£432.00	
	Interest	£ 34.56	8% of £432 = £34.56
	Total	£466.56	

Year 3	Capital	£466.56	
	Interest	£ 37.28	8% of £466 = £37.28
	Total	£504.40	

Total interest £504.40 − £400 = £104.40

The example on the previous page shows that by withdrawing the interest each year Sarah earned £96 simple interest after 3 years compared with £104.40 compound interest.

Exercise 4

Calculate the compound interest on the following investments.

1 £500 for 2 years at 6% pa.
2 £400 for 2 years at 8% pa.
3 £1500 for 3 years at 10% pa.
4 £2300 for 3 years at 4% pa.
5 £750 for 3 years at 6% pa.
6 £960 for 3 years at 4½% pa.

7 Liz invests £600 in a building society. She leaves her money invested for 3 years. In the first year the interest rate is 5%, in the second year the interest rate is 6% and in the third year the interest rate is 5½%. How much interest will she have earned?

Information ⇨   Banks and building societies add interest every 6 months.

Example   **Compound interest added twice per year**
If Sarah was paid interest on her £400.00 deposit every 6 months how would this affect her deposit after 1 year?

To find the first 6-monthly interest you find the annual interest then half it. The next 6-monthly interest is worked out as a percentage of the total amount after the first 6 months.

Period 1	Capital	£400.00	Annual interest = 8% of £400 = £32
	Interest	£ 16.00	½ of £32 = £16
	Total	£416.00	

Period 2	Capital	£416.00	Annual interest = 8% of £416 = £33.28
	Interest	£ 16.64	½ of £33.28 = £16.64
	Total	£432.64	

Total interest after 1 year = £32.64.

After 1 year she has earned an extra £0.64 interest (i.e. £32.64−32.00) by getting interest after each 6 month period rather than over 1 year.

Exercise 5

1   Continue the example to find how much extra Sarah gets after 3 years if she continues to get 6-monthly interest payments. Set your working out as shown above.

Calculate the compound interest on the following investments in questions **2-5**. In each case the interest rate quoted is per annum but the interest in added every 6 months.

2   £500 for 2 years at 6% pa.   3   £400 for 2 years at 8% pa.

4   £1500 for 3 years at 10% pa.   5   £2300 for 3 years at 4% pa.

6   Compare the answers to questions **2-5** with your answers to questions **1-4** in Exercise 4. What do you notice?

7   Tom won £750 at the Bingo. He decides to invest it in a building society where he will get 7½% interest. How much interest will he get if he leaves it for 3 years. (The interest is compound interest and is added every 6 months.)

I THINK I'D BETTER INVEST THIS...

Investigation   Is it better for an investor to have interest added as often or as seldom as possible?

Consider investing £100 at 8% pa for 2 years.
How much interest is earned if the interest is added annually?
How much interest is earned if it is added every 6 months?
How much interest is earned if it is added every 3 months?

Write down your conclusions.

Loans

Information ⇨ Banks give loans to customers. These loans may be used to buy a car, decorate a house, pay for a holiday ..... The amount the bank will lend depends on a number of things, for example, the amount you earn, the amount you spend each month, the reason for the loan etc. The bank will charge you **interest** on the loan, just as they pay you interest on deposits. The interest rate which they charge borrowers is much higher than the rate they pay to investors. So, it costs money to borrow money from banks.

Example

Using repayment tables for bank loans
The table opposite is used to calculate repayments for bank loans and is based on a loan of £100.

Term months	Monthly repayment
6	£17.62
12	£ 9.29
18	£ 6.51
24	£ 5.12
30	£ 4.29
36	£ 3.73
42	£ 3.33
48	£ 3.04
54	£ 2.81
60	£ 2.62

Anna is still thinking about buying the Fiesta. She has £500 saved up and would need a bank loan for £1500. Use the table to work out how much this would cost her per month if she took the loan out over 3 years?
How much interest is she paying?

3 years = 36 months
From the table, £100 for 36 months
= 3.73 per month

£1500 loan for 36 months
$£3.73 \times \dfrac{1500}{100} = £55.95$ per month

Anna would have to pay £55.95 per month for 3 years.
She would pay back a total of
£55.95 × 36 = £2014.20.
The interest she would pay is
£2014.20 − £1500 = £514.20.

Exercise 6

For each of these loans find (**a**) the monthly repayment and (**b**) the total interest paid.

1 Dave borrows £400 for 2 years to buy a stereo.

2 Najide borrows £700 for 1½ years to buy computer equipment.

3 Nelson borrows £3000 for 4 years to buy a sailing boat.

4 Sue borrows £750 for 1½ years to pay for a holiday.

5 Alan borrows £2450 for 3 years to pay for a fitted kitchen.

6 Mike borrows £5250 for 3 years to buy a new word processor.

7 Ian borrows £350 for 1 year to buy some new skiing equipment.

8 Stephen borrows £85 000 for 5 years to buy a small aeroplane.

APR

Information ⇨ **APR** stands for **Annual Percentage Rate** and it is a measure of different interest rates which are not calculated annually. The easiest way to calculate APR is to find out what would happen to £100 after 1 year.

Example

Calculating APR

If compound interest of 8% pa (per year) is added every 6 months, what is the APR?

Find out what would happen after 1 year if £100 was deposited at this rate.

Period 1	Capital	£100.00
	Interest	£ 4.00
	Total	£104.00

Period 2	Capital	£104.00
	Interest	£ 4.16
	Total	£108.16

Annual interest £108.16 − £100 = £8.16

So the APR = 8.16%

Investigation   Find out what the current interest rates are for investments in a number of banks and building societies. They often give a higher rate for large investments.

Calculate the APR for these interest rates.

Information ⇨ As well as giving interest when you invest money with them, banks and building societies charge you interest if you borrow money from them. Credit cards are another way in which interest is charged to borrow money.

Investigation   Find out how credit cards are used and how the goods or services are paid for

Most credit card companies charge interest at 2% per month. What is the APR?

Hire purchase

Information ⤸ Saving up to buy expensive articles like televisions, videos, stereos, cars, can take a long time. **Hire purchase** makes it possible to buy these things before you have saved up enough money to pay for the article in full. You usually have to pay a **deposit** and then sign an agreement to make regular payments to make up the remainder of the cost. The deposit is either given as a sum of money or as a percentage of the cash price. Normally the total **HP** cost is more than the cash price. Why do you think this is the case?

Example **Calculating the H P cost with deposit in £**
Look at the advert.

How much more will it cost to buy this video on hire purchase?

Total cost: £50+£24.43×12 = £343.16

Extra cost: £343.16−£325 = £18.16

Example **Calculating HP cost with percentage deposit**
A second-hand car costs £750. Under hire purchase a 20% deposit is paid followed by 18 monthly instalments of £37.45. What is the extra paid using HP?

Deposit: 20% of £750 = £150
HP cost: £150+£37.45×18 = £824.10
Extra cost: £824.10−£750 = £74.10

Exercise 7 Look at each of the adverts in questions **1-12** and work out how much extra is paid if the article is bought using hire purchase.

1 Television

2 Stereo

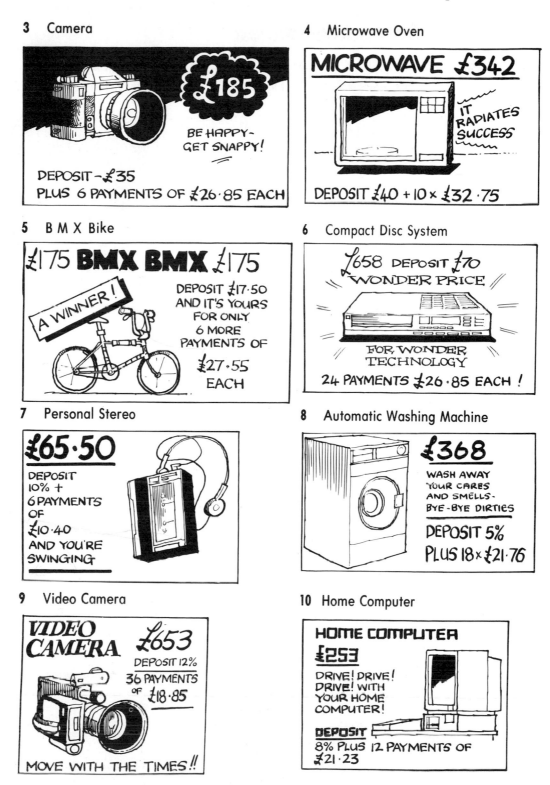

3 Camera

£185

BE HAPPY - GET SNAPPY!

DEPOSIT - £35
PLUS 6 PAYMENTS OF £26·85 EACH

4 Microwave Oven

MICROWAVE £342

IT RADIATES SUCCESS

DEPOSIT £40 + 10 × £32·75

5 B M X Bike

£175 BMX BMX £175

A WINNER!

DEPOSIT £17·50 AND IT'S YOURS FOR ONLY 6 MORE PAYMENTS OF £27·55 EACH

6 Compact Disc System

£658 DEPOSIT £70
WONDER PRICE

FOR WONDER TECHNOLOGY

24 PAYMENTS £26·85 EACH !

7 Personal Stereo

£65·50

DEPOSIT 10% + 6 PAYMENTS OF £10·40 AND YOU'RE SWINGING

8 Automatic Washing Machine

£368

WASH AWAY YOUR CARES AND SMELLS - BYE-BYE DIRTIES

DEPOSIT 5%
PLUS 18 × £21·76

9 Video Camera

VIDEO CAMERA £653

DEPOSIT 12%
36 PAYMENTS OF £18·85

MOVE WITH THE TIMES !!

10 Home Computer

HOME COMPUTER
£253

DRIVE! DRIVE! DRIVE! WITH YOUR HOME COMPUTER!

DEPOSIT
8% PLUS 12 PAYMENTS OF £21·23

11 Pam decides to buy a motor bike which is on sale for £865. She can also buy it under hire purchase by paying a 12% deposit followed by 24 monthly payments of £34.56. How much extra would she pay if she buys the motor bike using HP?

12 Gareth and Sarah have spent all their savings on carpets for their new flat. But they need something to sit on in the living room! They see a 3-piece suite on sale for £675 cash or under the following HP terms: no deposit and 15 monthly payments of £51.56. How much more will it cost them if they pay by HP?

Percentage interest

Information ⇨ The extra amount paid under hire purchase above the cash price is also called interest. This interest is usually expressed as a percentage of the original cash price.

Examples

Calculating percentage interest

1 A colour television costs £360. It can be bought by paying a deposit of £42 plus 12 monthly payments of £31.

    (a) How much interest is paid under HP?

    (b) Express the interest as a percentage of the cash price.

    (a) HP cost: £42+£31×12 = £414
        Interest: £414−£360 = £54

    (b) Percentage interest: $\dfrac{54}{360}$ = 15%

On a calculator 54 $\boxed{\div}$ 360 $\boxed{\%}$ 15

2 A home computer costs £380. It can also be bought by paying a deposit of £60 followed by 18 monthly payments of £19.50. Express the interest paid under hire purchase as a percentage of the cash price.
HP cost:　£60+£19.50×18 = £411
Interest: £411−£380 = £31

Percentage interest: $\dfrac{31}{380}$ = 8.2%

(On a calculator 31 $\boxed{\div}$ 380 $\boxed{\%}$ 8.1578974 rounded to 8.2%)

Exercise 8

Calculate the interest paid when buying each of the items in questions **1-5** using hire purchase. Express the interest as a percentage of the cash price in each case.

	Item	Cash price	HP terms Deposit	Payments
1	Home Computer	£360	£60	24 at £14
2	Food Processor	£120	–	15 at £10
3	Telephone	£80	£12	10 at £7.20
4	Skis	£90	£12	12 at £8
5	Ski Boots	£60	£13	10 at £5.60

HP terms

Information ✍   As you have seen, buying goods using hire purchase is usually more expensive than paying cash. Most shops have a standard HP arrangement such as 10% deposit, 15% interest. The HP terms for individual items are then worked out from this.

Example   **Calculating HP terms**
A shopkeeper advertises a video on sale for £376. It can also be bought on hire purchase over 1 year. The shopkeeper charges 15% interest for HP and asks for a deposit of 10% of the cash price. Decide on HP terms for the video.

HP interest:	15% of £376 = £56.40
Total HP cost:	£376+£56.40 = £432.40
HP deposit:	10% of £376 = £37.60
Total cost of instalments:	£432.40−£37.60 = £394.80
Cost of each instalment over 12 months:	£394.80÷12 = £32.90

Terms: Deposit £37.60 plus 12 monthly payments of £32.90

Example   **Calculating HP terms with rounding**
The shopkeeper has a vacuum cleaner on sale for £95. What hire purchase terms should he offer?

HP interest:	15% of £95 = £14.25
Total HP cost:	£95+£14.25 = £109.25
Deposit:	10% of £95 = £9.50
Total cost of instalments:	£109.25−£9.50 = £99.75
Cost of each instalment over 12 months:	£99.75÷12 = £8.3125

(Rounding off gives £8.31)

12 payments of £8.31 = £99.72

Total instalments = £99.75

The £0.03 difference is added to the deposit of £9.50. If no deposit is paid the difference Is added to the first instalment.

Terms: Deposit £9.53 plus 12 payments of £8.31

Exercise 9   Decide on the terms that the shopkeeper would charge for the following goods if his terms are 10% deposit and 15% interest over 12 payments. Set your working out like the examples above.

1	35 mm Camera	Price £180	2	Stereo	Price £360
3	Motor Bike	Price £720	4	Wardrobe	Price £124
5	Video Camera	Price £365	6	Golf Clubs	Price £473
7	Compact Disk	Price £567	8	Carpet	Price £289

Profit and loss

Information ➯ Industry buys raw materials and turns them into goods which are in demand. A **profit** is made if the goods are sold for more than it cost to make them, if not, a **loss** is made. The costs involved in making goods are more than just the raw materials. For example, in a factory, heating, lighting, wages, rent, rates, machinery all have to be paid for. These are called **overheads**.

In profit and loss problems there are 2 things to consider (a) the total cost price and (b) the total selling price. These 2 things must be calculated separately before the profit or loss can be determined.

Example   **Calculating profit or loss**

1   Bert has a stall at the local Saturday market which costs him £5 to rent. He buys 60 cassette tapes costing £28.20 per dozen and sells them at £7.65 for 3. Find his profit or loss when all 60 are sold.

Cost Price 60 (5 dozen)
$$= £28.20 \times 5$$
$$= £141.00$$
Plus rent of £5.00 = £146.00

Selling Price 60 (60 ÷ 3 = 20)
$$= £7.65 \times 20$$
$$= £153.00$$

Selling price is more than the cost price so Bert made a profit.

Profit £153.00 − £146.00 = £7.00

2   Another day he buys 60 screwdrivers for £25. He sells them at 3 for £1.30. Find his profit or loss.

Cost price of 60   = £25.00

Plus rent of £5.00 = £30.00

Selling price of 60 (60 ÷ 3 = 20)
$$= £1.30 \times 20$$
$$= £26.00$$

Selling price is less than the costs price so Bert made a loss.

Loss is £30 − £26 = £4.00

Exercise 10   Find Bert's profit or loss once each of the following have been sold in a day. Do not forget to include the rent.

1   100 video tapes costing £25.80 for 10 selling for £3.15 each.

2   48 RIP3 batteries costing £21.60 selling for 53p each.

3   10 radios costing £16.75 each selling for £18.99 each.

4   72 plugs costing £8.20 per dozen selling for £2.50 for 3.

5   100 computer disks costing £15.45 for 10 selling for £2.45 each.

6   20 personal stereos costing £385 selling for £19.99 each.

7   180 light bulbs costing £3.20 for 10 selling for £1.05 for 3.

8   10 tool sets costing £72.45 selling for £7.25 each.

9   Bert buys screwdrivers of 3 different sizes and makes them up into kits which he sells for £2.99 each. He buys the screwdrivers in packs of 10 costing £7.45, £8.50 and £10.85 respectively. What is his profit or loss on each kit?

Information ➯   Sometimes we need to compare profits made from selling one sort of item against profits from another. Profits are difficult to compare if the items have different cost prices. In each case we can work out the profit as a percentage of the cost price then compare the 2 percentages. This is known as calculating the **percentage profit**.

Example   **Calculating percentage profit or loss**
In example 1 on the previous page you found that Bert made a profit of £7 when he sold all the cassettes. The cassettes cost him £141. What was his percentage profit?

Percentage profit = $7 \div 141\% = 5\%$

Exercise 11   For questions **1-9** find the percentage profit or loss which Bert makes on each of the 9 items in exercise 10.

10   Which of the 9 items gives Bert the best profit?

11   On which of the 9 items does Bert make the biggest loss?

Price fixing

Information ➪ Shopkeepers have to decide on the selling price of each item they sell. If they charge too much, they will lose custom. If they do not charge enough, they will make a loss and go out of business.

Example

Fixing a price
If a shopkeeper pays £125 for a box of 50 calculators, what should be the lowest selling price for each calculator?

Cost per calculator £125÷50 = £2.50
If the shopkeeper only charges £2.50 she will not cover her overheads.
If she charges too much they will not sell, so she may decide to charge about £2.75.

Exercise 12

In questions **1-5** below the cost prices of various goods from a school tuckshop are given.

(a) How much would you charge for each item ?

(b) Work out the percentage profit for each item based on your selling price.

1   48 bags of crisps costing £7.80

2   72 bars of chocolate costing £11.50

3   60 bottles of lemonade costing £11.15

4   144 bags of sweets costing £13.50

5   48 fruit bars costing £9.80

6   A crate containing 24 two litre bottles of lemonade costs £19.44. Glasses of lemonade are sold in 200 ml measures. How much should a cafe owner charge for each glass to make at least a 15% profit?

7   You are organising a disco to raise funds for a local charity. You expect to sell 250 tickets. Here are the costs.

> Hire of hall £45
> Hire of disco £65
> Hire of lights.............. £35
> Printing tickets £15

(a) Decide on the price for a ticket.

(b) Work out how much you should be able to donate if all 250 tickets are sold.

(c) How many tickets must be sold to break even ?

8   In a delicatessen, the owner makes up her own blend of tea. She mixes 5 lb of tea A, 3 lb of tea B and 8 lb of tea C. The teas cost her £2.41 per lb, £2.89 per lb and £3.09 per lb respectively.

(a) Find the cost price of one quarter pound of tea.

(b) Decide on a selling price and find the profit per quarter of tea.

(c) Find the percentage profit correct to 1 decimal place.

Buying a car

You will now realise that there is no single answer to Anna's problem.
There are various choices open to her. She will need to find out information on interest rates for different loans and then work out the cost of each option. The Garage HP deal is worked out here for you. You should collect information on the other forms of payment and make a comparison yourself. Think about the following questions.

Applying for a bank loan
How much does Anna need to borrow?
What would the repayments be for different periods of loan?
How much interest is charged?

Saving up for enough money to buy car
How quickly does Anna want or need the car?
Would she be better using what she can save to pay off a bank loan?

Buying the car using hire purchase
The garage offered 25% deposit and 5% interest.

Cost £2000. Deposit is 25% of £2000 = £500
Amount Anna has to pay interest on is £2000−£500 = £1500
Interest is 5% of £1500 = £75.

Using a credit card to buy the car
Is it within her credit limit?
How long does she intend to take to repay the loan?
What will her interest charges be over this time?

Investigations

1   Choose an item from a mail order catalogue and work out how you would buy it.

2   Pay a visit to a local electrical shop. Choose an item which costs £300 to £500. Find out the HP terms on offer. Compare this with the cost of a bank loan. Which method of payment is best?

3   Robert's old washing machine has broken down and he has been told that it is not worth repairing. He has savings of £120 and can afford to put £10 per week towards a new washing machine. Go to your local electrical shop and decide which washing machine you would advise him to buy. How should he pay for it?

8 DIY

Do you remember? ⇔ In this unit it is assumed that you know how to:

1 multiply numbers

2 divide numbers

3 do calculations involving a mixture of operations

4 Round off answers to a given number of significant figures.

Exercise 0 will help you check. Can you answer all the questions?

Exercise 0

Multiplying numbers
Calculate the answers to

1 24×23	2 12×45	3 18×18
4 2.4×3.6	5 5.1×3.2	6 2.7×4.2
7 5.2×3.4	8 2.4×6.2	9 3.8×3.2

Dividing numbers
Calculate the answers to

10 36÷12	11 45÷13	12 63÷17
13 36÷12.5	14 145÷14.6	15 321÷10.5
16 23.45÷8.4	17 143÷23.2	18 235.63÷12.6

Mixed operations
Calculate the answers to

19 6×2+4×3	20 3×12+4×8
21 12×4−3×8	22 45×2−12×2
23 34÷2+16÷4	24 24÷8+12÷3
25 2+15×3+8	26 23+2×4−8
27 3×5+2×4+6×2	28 8×2.5+4×3.6+2×1.7

Rounding
Express these numbers correct to 3 significant figures.

29 588.2	30 480.7	31 338.5
32 36.27	33 29.44	34 62.68
35 4.662	36 3.448	37 2.465
38 3668	39 28569	40 242678

Fire damage

Peter came home late one night and decided to make himself some chips. Unfortunately he dozed off to sleep as the oil was heating.........

The fire was quickly brought under control. There was no structural damage although his kitchen and living room now need redecorating due to smoke damage. Peter likes decorating and decides to redecorate the rooms himself because he cannot trust anybody else to do it as well! He needs to give the insurance company a detailed estimate so he can claim the cost of the job from them.

On the right are the floor plans for the 2 rooms which are both 2.6 metres high. Peter estimates the total area of woodwork (the skirting boards) to be 7 m².
The walls, ceiling and woodwork need painting in the kitchen, the ceiling and woodwork need painting in the living room and the living room walls need papering. How much will the materials for the whole job cost?

Below is some of the information Peter has gathered from his local DIY store.

Wallpaper £4.35 per roll

Paste £1.69 for up to 8 rolls

Emulsion paint: 1 litre covers 11 m²
     £3.95 for 1 litre tin
     £7.25 for 2½ litre tin

Gloss paint: 1 litre covers 15 m²
     £2.29 for ½ litre tin
     £3.95 for 1 litre tin
     £7.25 for 2½ litre tin

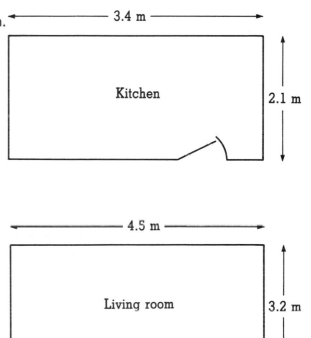

This unit will help you estimate the cost of decorating jobs like this.

Areas

Information ⤸ You need to cover a floor with square tiles — how many should you buy?
You want to paint a room — how much paint will you need?
You want to sow a lawn — how many packets of seed will you need?
To answer these questions you need to know the **areas** you have to cover. You should already know some some of the common units of area.

Examples

Units of area
A **pencil point** has an area of about
1 square millimetre (1 mm²).

That is an area equal to a square
1 mm by 1 mm.

A **typewriter key** has an area of about
1 square centimetre (1 cm²).

That is equal to a square
1 cm by 1 cm.

A **large window** has an area of about
1 square metre (1 m²).

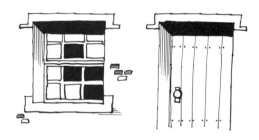

That is equal to a square
1 m by 1 m.

A **large village** has an area of about
1 square kilometre (1 km²).

That is equal to a square
1 km by 1 km.

A **page in this book** is about 540 cm².

That means that the page could be
covered by about 540 centimetre
squares.

Exercise 1

Use a suitable unit from the 4 examples above and guess the area of the following.

1 A room floor	2 A bedroom wall	3 A field
4 A small garden	5 A stamp	6 A page
7 A school desk	8 A door	9 A 10p piece
10 A small forest	11 A large city	12 A swimming pool

Basic shapes

Information    Here are 2 basic shapes whose area you should know how to find.

Area of a square $A = l^2$ where l is the length of the side.
Area of a rectangle $A = lb$ where l and b are the length and breadth.

Example   **Area of a square**

A draughts board is 30 cm square.
What is its area?

Area of a square $A = l^2$
                   $l = 30$ cm
                   $A = 30^2$ cm²
                   $A = 900$ cm²

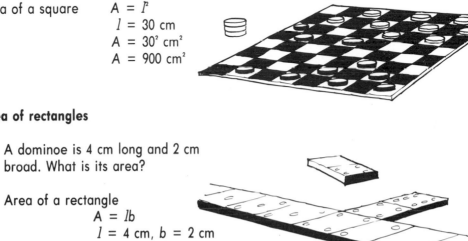

Examples   **Area of rectangles**

1   A dominoe is 4 cm long and 2 cm
broad. What is its area?

Area of a rectangle
         $A = lb$
         $l = 4$ cm, $b = 2$ cm
         $A = 4 \times 2$ cm²
         $A = 8$ cm²

2   A rectangular field is 45 m long and
32 m broad. What is its area?

Area of a rectangle
         $A = lb$
         $l = 45$ m, $b = 32$ m
         $A = 45 \times 32$ m²
         $A = 1440$ m²

Exercise 2   Find the area of each of these squares or rectangles.

1   A rectanguler field which is 150 m long by 90 m broad.

2   The floor of a room which is 6 m long by 4 m broad.

3   A games board which is a 40 cm square.

4   A stamp which is 30 mm long by 20 mm broad.

5   A wall which is 6.5 m long by 2.7 m high.

6   A page of a book which is 20 cm broad by 30 cm long.

The triangle

Investigation

1   On 1 cm squared paper draw 5 rectangles each with different lengths and breadths and find the area of each rectangle.

2   Draw one diagonal in each rectangle.

3   Cut out the rectangles and cut each one along the diagonal.

4   With each rectangle try to fit one triangle on top of the other. Do they always fit?

6   Draw a rectangle and mark a triangle inside as shown in the diagram.

7   Cut it out and try to fit the 2 white triangles on top of the shaded one.

The area of a triangle is half the area of the surrounding rectangle.

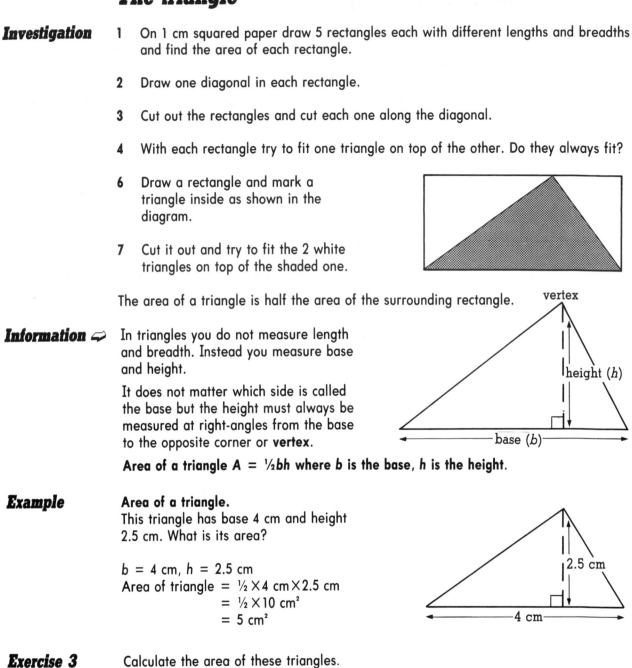

Information ✑   In triangles you do not measure length and breadth. Instead you measure base and height.

It does not matter which side is called the base but the height must always be measured at right-angles from the base to the opposite corner or **vertex**.

Area of a triangle A = ½bh where b is the base, h is the height.

Example

Area of a triangle.
This triangle has base 4 cm and height 2.5 cm. What is its area?

b = 4 cm, h = 2.5 cm
Area of triangle = ½ × 4 cm × 2.5 cm
  = ½ × 10 cm²
  = 5 cm²

Exercise 3   Calculate the area of these triangles.

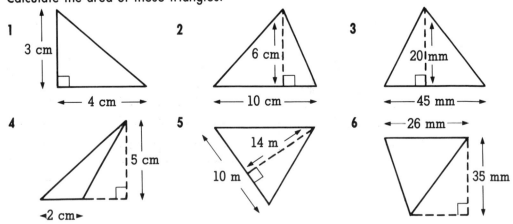

1   3 cm   ← 4 cm →

2   6 cm   ← 10 cm →

3   20 mm   ← 45 mm →

4   5 cm   ◄2 cm►

5   14 m   10 m

6   ← 26 mm →   35 mm

Transforming shapes

Information ➾ In geometry shapes can be moved or changed in different ways. These are called **transformations**. The new shape in each case is called the **image**. Here are 4 important types of transformations.

Examples

Reflection	Rotation (+90°)	Translation	Similarity (Scale factor 2)

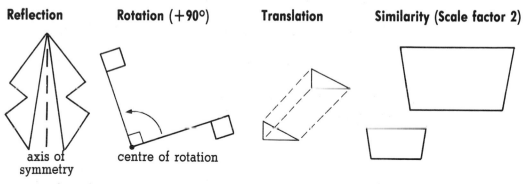

axis of symmetry   centre of rotation

Exercise 4 Copy each of the following shapes onto squared paper and draw the image under the given transformation.

1 Reflection: axis of of symmetry AB

2 (a) 90° rotation about O
 (b) Similarity: scale factor 2

3 Translation: 3 squares right and 1 square up

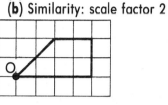

4(a) Similarity: scale factor 3
 (b) Rotation 180° about 0

5 Reflection in PQ

6 Translation: 2 squares left and 3 squares down

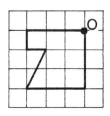 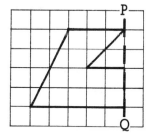 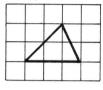

7 Reflection: axis of symmetry XY

8 180° rotation about O

9 (a) Similarity: scale factor ½
 (b) Reflection in AB

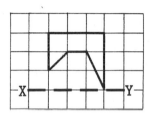 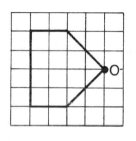 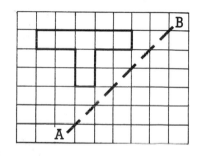

Information ⇨ Many common shapes are **symmetrical**. This means that you can draw 1 or more **axes of symmetry** through the shape. To identify an axis of symmetry think about folding the shape along the axis so that each half fits exactly onto the other.

You can also identify the **centre of symmetry** in some shapes. To find the centre of symmetry try to find a point about which the shape can be turned to fit its own outline.

Example   **Identifying symmetry**
Draw the axes of symmetry and mark the centre of symmetry in these shapes.

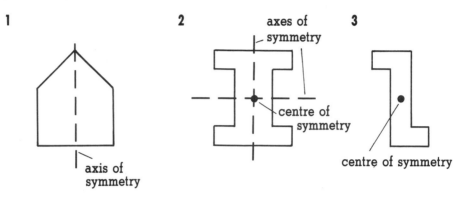

Exercise 5   Copy each of these shapes onto squared paper, draw in any axes of symmetry and mark in the centre of symmetry where appropriate.

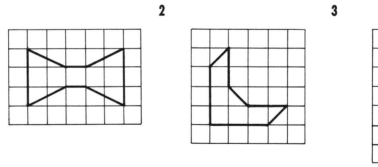

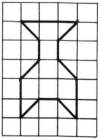

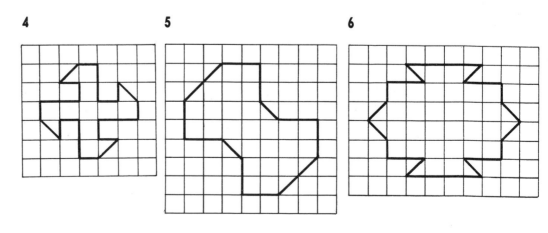

Other familiar shapes

Information ⇨ Here are 3 other shapes which can be made up from triangles through the transformations given.

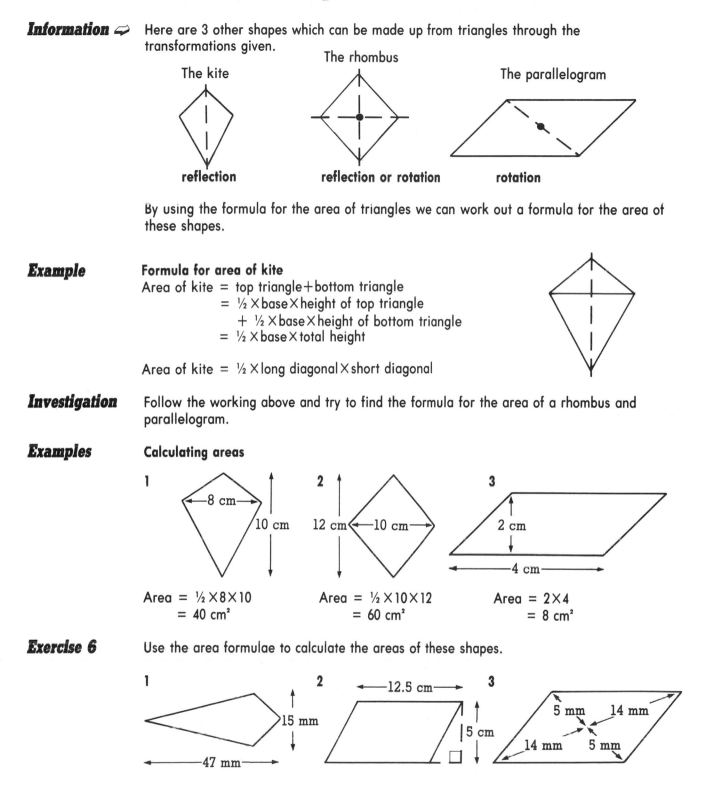

The kite

The rhombus

The parallelogram

reflection   **reflection or rotation**   **rotation**

By using the formula for the area of triangles we can work out a formula for the area of these shapes.

Example   **Formula for area of kite**
Area of kite = top triangle+bottom triangle
= ½ ×base×height of top triangle
+ ½ ×base×height of bottom triangle
= ½ ×base×total height

Area of kite = ½ ×long diagonal×short diagonal

Investigation   Follow the working above and try to find the formula for the area of a rhombus and parallelogram.

Examples   **Calculating areas**

1
←8 cm→
10 cm

Area = ½ ×8×10
= 40 cm²

2
12 cm ←10 cm→

Area = ½ ×10×12
= 60 cm²

3
2 cm
←4 cm→

Area = 2×4
= 8 cm²

Exercise 6   Use the area formulae to calculate the areas of these shapes.

1
15 mm
←47 mm→

2
←12.5 cm→
5 cm

3
5 mm   14 mm
14 mm   5 mm

The circle

Information ⇨ You will often need to be able to calculate circular areas. The area of a circle is found from the formula $A = \pi r^2$ where r is the radius of the circle and the value of π is 3.14 correct to 3 significant figures. Answers to problems which use 3.14 for π should be given correct to 3 significant figures. The symbol π is the greek letter 'pi'.

Example

Area of a circle

1 What is the area of the circle shown in the diagram?

   Area $= \pi r^2$ and $r = 5$ cm
   Area $= 3.14 \times 5^2$
   $\quad = 3.14 \times 5 \times 5$
   $\quad = 78.5$ cm^2

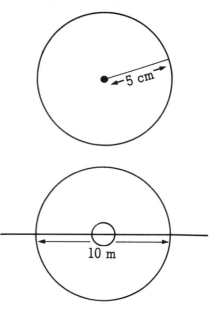

2 The centre circle of a football pitch has a diameter of 10 m. What is its area?

   Area $= \pi r^2$ and $d = 10$ so $r = 5$
   Area $= 3.14 \times 5^2$
   $\quad = 3.14 \times 5 \times 5$
   $\quad = 78.5$ m^2

Exercise 7

Use the formula $A = \pi r^2$ to solve these problems.

1 Find the area of a circle of (**a**) radius 7 cm (**b**) radius 15 m (**c**) diameter 18 mm (**d**) diameter 35 cm (**e**) radius 27 mm

2 Angela has a circular plot of land in her garden which she wants to sow with grass seed. She measures the diameter of the plot to be 3 m. What area of ground is this?

3 A 10p piece has a diameter of 28 mm. What is its area?

4 A record has a radius of 9 cm. What is its area?

5 Copy and complete this table.

	Radius	Diameter	Area
(a)	4 cm		$3.14 \times 4 \times 4 = 50.2$ cm^2
(b)	8 cm		
(c)		12 mm	
(d)		24 mm	

What happens to the area of a circle if the radius is doubled?

Information ⇨ Sometimes when you do DIY work you need to find out the distance round circular objects. For example you may need to put a strip of edging round a circular coffee table. The distance round a circle is called **the circumference**. The formula for calculating this is **c** = π**d** where c is the circumference, π = 3.14 to 3 significant figures and d is the diameter.

Example

Calculating the circumference

1   Here is a 1p coin. Its diameter is 2 cm. What is its circumference?

$c = \pi d$ and $d = 2$ cm
$c = 3.14 \times 2$
    $= 6.28$ cm

2   The radius of a 10p piece is 14 mm. What is its circumference?

$c = \pi d$ and ($r = 14$ mm so $d = 28$ mm)
$c = 3.14 \times 28$
    $= 87.9$ mm

Exercise 8

Use the formula $c = \pi d$ to solve these problems.

1   Tony has made a circular coffee table of radius 55 cm. He wants to buy an adhesive strip to edge it. What length should he buy?

2   Gordon wants to build a fence round his circular fish pond. The builder's merchant needs to know the length of fence so that he can deliver the correct amount of wood. Gordon's pond is 3 m in diameter. What will be the length of the fence needed?

3   Measure the diameter of a £1 coin, and then work out its circumference. How could you check your answer.

4   Copy and complete this table

	Radius	Diameter	Circumference
(a)	4 cm		
(b)	8 cm		
(c)		20 mm	
(d)		40 mm	

(e)   Look at your answers to the table. When the radius is doubled what happens to the circumference?

Composite shapes

Information ➷ **Composite shapes** are shapes which are made up from 2 or more common shapes. To find the area of these shapes it is best to split them up into simple shapes.

Example   **Calculating the area of composite shapes**
Here is an arched window. What area of glass is needed for the window?

Divide the shape into a rectangle and a semi-circle. Calculate the areas.

Rectangle: $A = lb$ $l = 1.5$ m $b = 1.2$ m
   $= 1.5 \times 1.2$
   $= 1.80$ m²

Semi-circle: $A = \frac{1}{2}\pi r^2$ $\pi = 3.14$ $r = 0.6$ m
   $= \frac{1}{2} \times 3.14 \times 0.6 \times 0.6$
   $= 0.565$ m²

Total area: $1.80 + 0.565 = 2.37$ m²

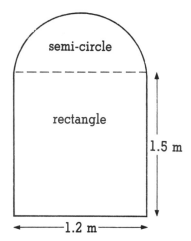

Exercise 9   Find the area of the shapes in the following problems by splitting them up into simple shapes.

1   Here is the plan of Judy's living room floor. What area of carpet would be required to cover it?

2   Keith wants to paint the side of his lean-to shed. He knows he has enough paint left to cover 8 m². Will he have enough for the side of his shed?

3   Bahi wants to buy vinyl tiles for his kitchen. Here is a plan of the floor. What is the total area to be tiled?

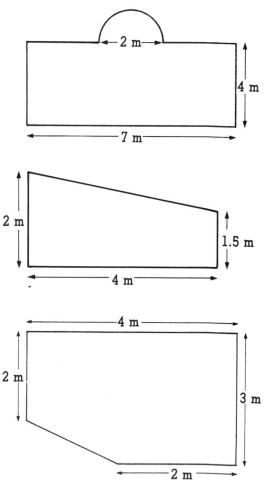

Investigation   Similar areas

Wes and Denzil are keen snooker players. They each have a small sized snooker table at home which they practise on. At the snooker club they play on full size tables.

Wes's table measures 3' by 1½'.

Denzil's table measures 6' by 3'.

A full size table measures 12' by 6'.

Consider the length and breadth of each table. Copy and complete:

1   The full size table is ☐ times larger than Wes's.

2   The full size table is ☐ times larger than Denzil's.

Calculate the area of each table. Copy and complete:

3   The area of the full size table is ☐ times the area of Wes's table.

4   The area of the full size table is ☐ times the area of Denzil's table.

Compare your answers to questions **1-4** above. Does the same scale factor apply to questions **1** and **2** as to questions **3** and **4**?

5   Morag has a snooker table measuring 4' by 2'. How many times larger than her table is the length and breadth of the full size table? How many times larger is the area of the full size table?

6   Copy and complete the table below.

	Scale factor for length & breadth	Scale factor for area
Wes Denzil Morag	4	16

What is the relationship between the scale factor for length and the scale factor for area?

Painting

In general, emulsion paint is used on walls and ceilings and undercoat and gloss paint are used on wood. When decorating you should be able to work out roughly how much paint you are going to need and how much it will cost. You usually need to use two coats of emulsion and gloss paint. The manufacturers of paint usually specify the area that 1 litre of paint will cover.

Example

Estimating the amount of paint
Find the amount of paint needed to cover the walls of the room below with 2 coats of emulsion. The door and the window are not going to be painted.

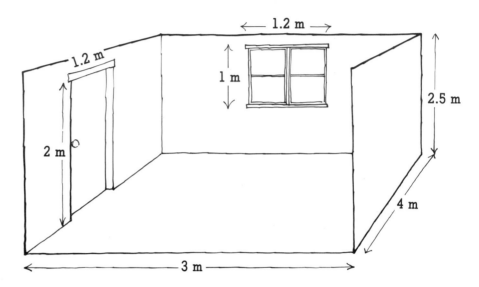

The area of each wall is found by using the formula $A = lb$.

Total area is $3 \times 2.5 + 4 \times 2.5 + 3 \times 2.5 + 4 \times 2.5 = 35$ m²

Area of door: $1.2 \times 2 = 2.4$ m²
Area of window: $1.2 \times 1 = 1.2$ m²

Area to be painted: $35 - 2.4 - 1.2 = 31.4$ m²

1 litre of emulsion paint will cover 10 m²
Paint needed for each coat $31.4 \div 10 = 3.14$ litres
For 2 coats you will need $2 \times 3.14 = 6.28$ litres

Emulsion paint is usually only sold in 1 litre or 2½ litre tins and costs £3.59 and £7.25 respectively. So there are a number of different ways to buy 6.28 litres of paint. For example, you could buy

(a) $2 \times 2½$ litre tins + 2×1 litre tins (total 7 litres).   Total cost = £21.68
or
(b) $3 \times 2½$ litre tins (total 7½ litres).   Total cost = £21.75
or
(c) $1 \times 2½$ litre tin and 4×1 litre tins (total 6½ litres).   Total cost = £21.61

Can you find other ways? Which is the cheapest?

Exercise 10

1 Copy and complete the table below. Each litre of emulsion paint covers an area of 10 m². Paint is available in 1 litre or 2½ litre tins costing £3.59 and £7.25 respectively. Try to find the cheapest way to buy the paint.

	Area to be painted	Amount of paint needed (litres)	Tins of paint needed	Cost
	50 m²	5 l	2 × 2½ l	£14.50
	35 m²	3.5 l	1 × 2½ l + 1 × 1 l	£10.84
(a)	25 m²			
(b)	60 m²			
(c)	45 m²			
(d)	70 m²			
(e)	30 m²			

2 Megan and Rhonda are painting their newly rented flat. Calculate how much it will cost them to paint the walls (but not the door or window) of the room shown below with 2 coats of emulsion paint. The paint is available in 1 litre and 2½ litre tins costing £3.62 and £7.28 respectively. One litre of the paint covers 12 m².

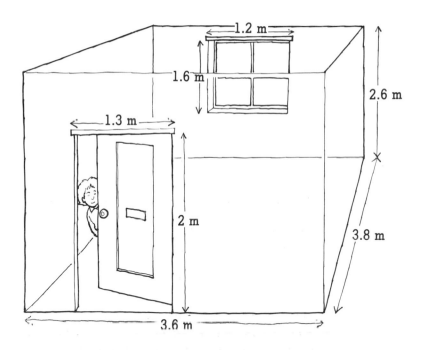

3 Calculate the cost of painting the door, skirting board and window frame of Megan and Rhonda's living room with one coat of gloss paint. (Assume that undercoat is not needed). The skirting board is 12 cm high and runs right round the room. The window frame is made from wood 11 cm wide. The paint is available in ½, 1 and 2½ litre tins costing £2.29, £3.95 and £6.29 respectively. One litre of gloss paint covers 12 m².

4 Measure the walls of your classroom. Calculate the area that needs to be painted with emulsion. For doors and woodwork calculate the area that needs to be painted with gloss. Use the prices from questions **2** and **3** to find the cost of painting your classroom? (Assume undercoat is not needed on the woodwork.) Could you estimate how much it would cost to paint the whole school?

Wallpaper

Information ↩ The chart below helps to calculate the number of rolls of wallpaper needed to paper different rooms. Only 2 measurements are required. These are the height to be papered and the perimeter of the room. Both are measured in metres. To find the perimeter of the room measure the length and breadth of the room and use the formula $P = 2(l+b)$. Do not subtract doors and windows as this allows some spare paper for matching up patterns.

Wallpaper chart

		PERIMETER (m)										
		10	11	12	13	14	15	16	17	18	19	20
H	2.0	4	4	5	5	5	6	6	6	7	7	8
E	2.2	4	5	5	5	6	6	7	7	7	8	8
I	2.4	5	5	5	6	6	7	7	8	8	9	9
G	2.6	5	5	6	6	7	7	8	8	9	9	10
H	2.8	5	6	6	7	7	8	8	9	10	10	11
T	3.0	6	6	7	7	8	8	9	10	10	11	11
(m)	3.2	6	7	7	8	8	9	**10**	10	11	11	12
	3.4	6	7	8	8	9	10	10	11	12	12	13

Example

Calculating the number of rolls
Find the number of rolls of wallpaper needed to decorate a room measuring 3.6 m broad by 4.2 m long by 3.2 m high.

Formula: $P = 2(l+b)$ and $l = 4.2$ m $b = 3.6$ m
Substituting: $P = 2 \times (4.2+3.6)$
$P = 2 \times 7.8$
$P = 15.6$ m

Always round up the perimeter of the room before using the table. You will always need *more* rather than *less* paper.

Perimeter = 15.6 m so look up perimeter of 16 m and height of 3.2 m.
From the table 10 rolls of wallpaper are needed.

Exercise 11   Use the wallpaper table to calculate the number of rolls of wallpaper needed for the following situations.

1   Andrew wants to paper his living room which is 3.8 m long, 3.4 m broad and 2.7 m high.

2   Dawn would like to wallpaper her bedroom which is 4.1 m long, 3.4 m broad and 3 m high.

3   Rahji is decorating the rooms in her ground floor flat. The ceilings are all 2.7 m high. How many rolls of paper would she need for each of the rooms?

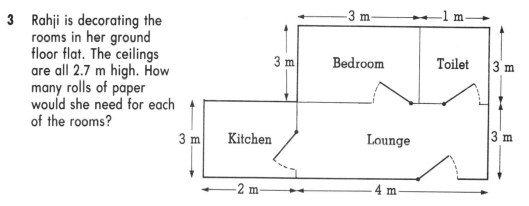

4   Peter has 6 rolls of wallpaper. His rooms are 2.4 m high. Which of these rooms could he use the wallpaper to decorate completely?

Kitchen   length 2.8 m   breadth 2.4 m
Lounge   length 4.3 m   breadth 3.6 m
Bedroom   length 3.2 m   breadth 2.7 m
Bathroom   length 3.4 m   breadth 2.3 m

5   Lloyd sees woodchip wallpaper offered at £0.55 per roll. How much will it cost him to decorate the rooms of his flat shown here. All his ceilings are 3.1 m high.

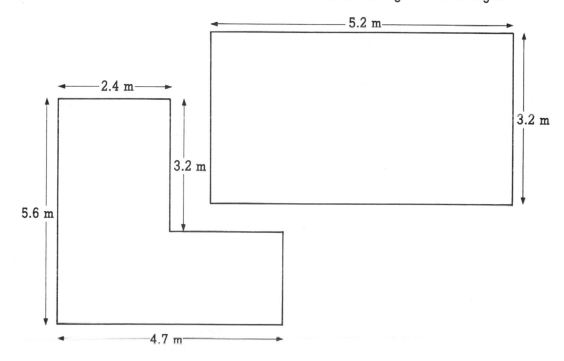

Floor coverings

Information ☞ Carpets and vinyl floor covering are sold in standard widths. Vinyl floor covering is sold in lengths either 2 metres wide or 3 metres wide. The cost of the floor covering is quoted per linear foot. You should always try to lay vinyl in one piece and this is important to remember when you decide which width to buy. Here is a metric to imperial conversion chart to help you work out the lengths to buy.

Metric (m)	0.1	0.2	0.3	0.4	0.5	0.6	0.7	0.8	0.9
Imperial	4"	8"	1'	1'4"	1'8"	2'	2'4"	2'7"	2'11"

Metric (m)	1.0	2.0	3.0	4.0	5.0				
Imperial	3'3"	6'7"	9'10"	13'1"	16'5"				

Remember 12 inches (12") = 1 foot (1')

Example

Cost of floor covering
Find the cost of covering the floor shown below in vinyl costing £6.99 per foot for a 3 m width.

Width of room = 2.8 m
Vinyl width = 3 m (more than enough)

You therefore need a piece of vinyl 3.8 m long.

From conversion chart
3.0 m = 9' 10"
0.8 m = 2' 7"

3.8 m = 11' 17" (17" = 1' 5")
  = 12' 5"

You need to buy 13' of 3 m vinyl.
Cost = £6.99 x 13
  = £90.87

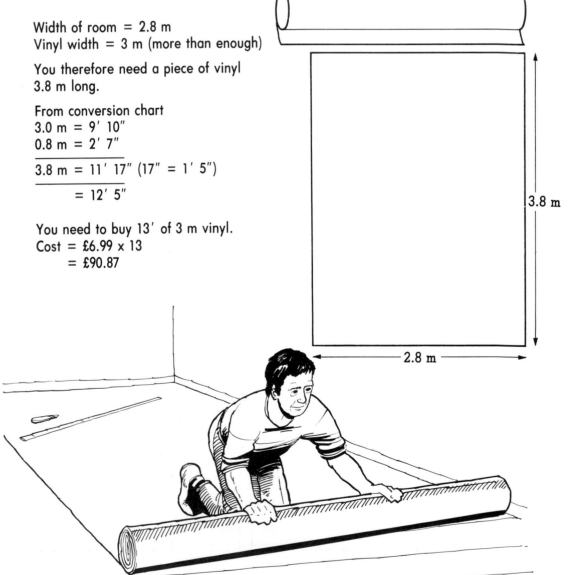

Exercise 12   Select the best width of vinyl to cover the floors shown below. Work out how many feet you need to buy and the total cost.
2 m wide vinyl floor covering costs £5.25 per linear foot
3 m wide vinyl floor covering costs £6.99 per linear foot

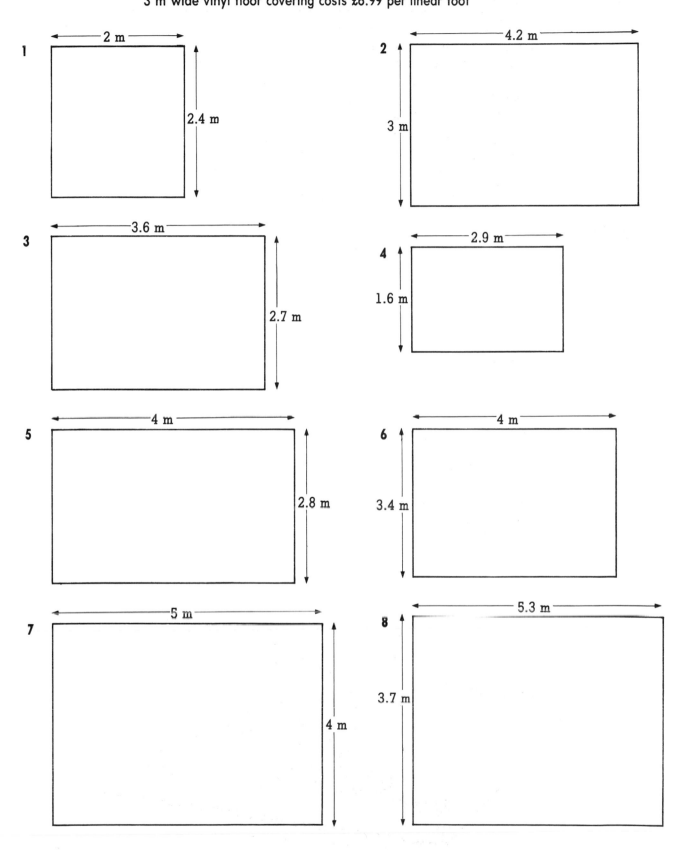

Volume

Information ➩ **Volume** is a measure of size. Here is a cuboid drawn on isometric dot paper. A **unit cube** is drawn beside it.

The volume of the cuboid is found by calculating the number of unit cubes that will fit inside the cuboid.

4 cubes will fit along the length.
3 cubes will fit along the breadth.
2 cubes will fit against the height.

The volume is $4 \times 3 \times 2 = 24$

If the measurements are made in centimetres then the volume is measured in cubic centimetres (cm^3).

The formula to find the volume of a cuboid is $V = lbh$
where v is the volume
  l is the length
  b is the breadth
  h is the height of the cuboid

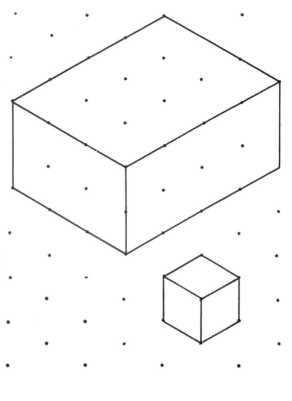

Example   **Calculating volume**
Find the volume of a cuboid measuring 1.2 m by 45 cm by 35 cm.

Convert all measurements to the same units (either metres or centimetres) before working out the volume.

Working in metres: Working in centimetres:
$V = lbh$ $V = lbh$
$V = 1.2 \times 0.45 \times 0.35$ or $V = 120 \times 45 \times 35$
$V = 0.189$ m³ $V = 189\,000$ cm³

Exercise 13   Using the formula find the volume of the following cubes and cuboids in questions **1-5**.

	Length	Breadth	Height	Volume
	8 cm	5 cm	6 cm	240 cm³
	1.5 m	40 cm	40 cm	0.24 m³ or 240 000 cm³
1	10 cm	5 cm	6 cm	
2	3 m	2 m	1.5 m	
3	1.5 m	60 cm	40 cm	
4	45 cm	45 cm	1.3 m	

5   A packet of washing powder measures 36 cm by 27 cm by 12 cm. What is its volume?

6   Measure the height, width and depth of this textbook. What is its volume?

7   A box of tissues measures 25 cm by 12 cm by 7 cm. What is its volume?

Prisms

Information ➥ The cross-section of this box is a triangle. The cross-section is the same at both ends, in the centre, or anywhere in between. A solid which has a constant cross-sectional area along its length is called a **prism**.

To find the volume of a prism multiply the cross-sectional area by the length.

Example   **Calculating the volume of a prism**

1 Find the volume of the tent shown in the sketch.

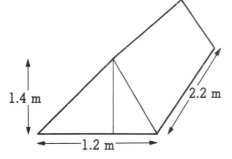

The cross section is a triangle with base
1.2 m and height 1.4 m.
Area = ½*bh*
Volume = area×length
$V = ½bh×l$
where $b = 1.2$ $h = 1.4$ $l = 2.2$
$V = 0.5×1.2×1.4×2.2$
$V = 1.848$ m³

2 A soup tin is a cylinder radius 3.5 cm and height 10.5 cm. What volume of soup does it contain?

A cylinder is a prism with a circular base. Area of base $= πr^2$
Volume = area×length so $V = πr^2×l$ where $π = 3.14$ $r = 3.5$ cm $l = 10.5$ cm
$V = 3.14×3.5×3.5×10.5$
$V = 404$ cm³

Exercise 14   Find the volume of the following triangular prisms and cylinders.

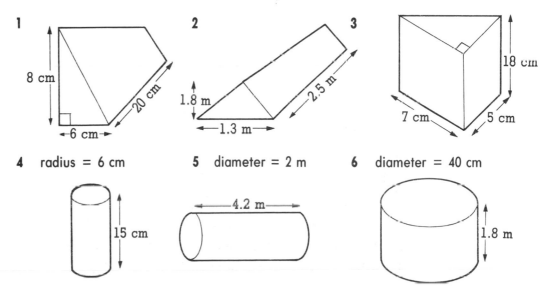

1   8 cm   20 cm   6 cm

2   1.8 m   2.5 m   1.3 m

3   18 cm   7 cm   5 cm

4 radius = 6 cm   15 cm

5 diameter = 2 m   4.2 m

6 diameter = 40 cm   1.8 m

Information ⤸ The exercise below contains a mixture of volume problems. You should choose the appropriate formula from the ones you now know.

Example **Calculating volume**
A cylindrical coffee mug is 8 cm high with a diameter of 6 cm. What is its volume? (Do not count the handle.)

Volume = Area of circle×height
Volume = $\pi r^2 h$
$\quad$ = $3.14 \times 3^2 \times 8$
$\quad$ = $3.14 \times 9 \times 8$
$\quad$ = 226.08 cm³

Exercise 15 **1** On isometric dot paper draw as many cuboids as you can with volumes of
(a) 4 cm³ (b) 6 cm³ (c) 12 cm³

2 Calculate the volume of a water tank which measures 150 cm by 100 cm by 85 cm.

3 A central heating system requires a boiler with a capacity of at least 120 000 cm³ Which of these boilers would be suitable for the system?

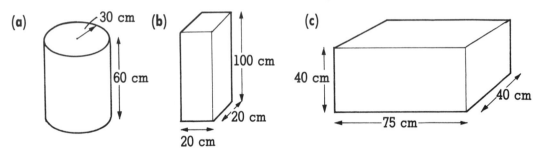

(a) 30 cm, 60 cm

(b) 100 cm, 20 cm, 20 cm

(c) 40 cm, 40 cm, 75 cm

4 Jason decides he has room for a goldfish pond in his garden. He chooses a ready made, circular, plastic mould for the pond which has diameter 2 m and depth 60 cm. What volume of water will be required to fill it?

5 The diagram shows a patio with a gravel path round three sides.
The gravel is 5 cm deep. What is the volume of gravel in the path?

6 Crown matchpots — small sample tubs — contain 75 ml of paint. How many of these would be equivalent to the large 2½ litre tin shown?

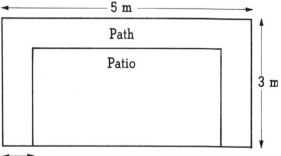

5 m
Path
Patio
3 m
50 cm

Fire damage

Here are the floor plans of the 2 rooms which Peter has to decorate. The walls are 2.6 metres high.

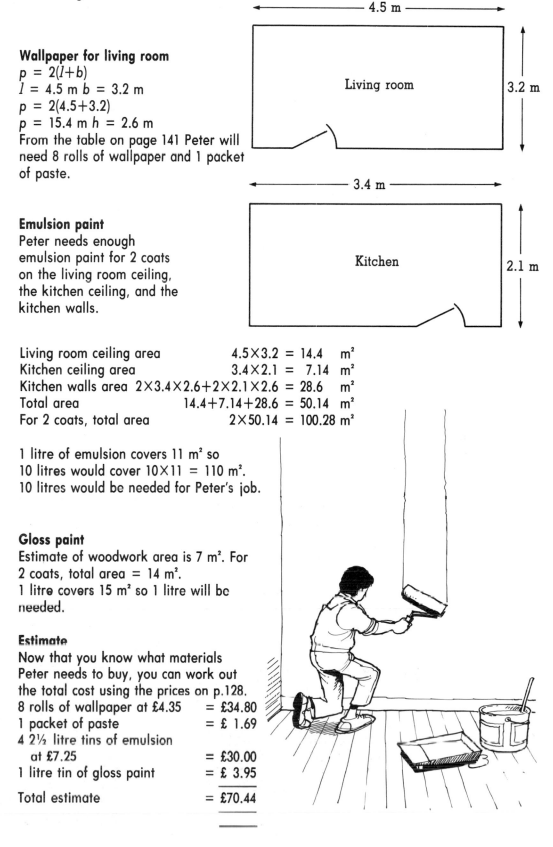

Wallpaper for living room
$p = 2(l+b)$
$l = 4.5$ m $b = 3.2$ m
$p = 2(4.5+3.2)$
$p = 15.4$ m $h = 2.6$ m
From the table on page 141 Peter will need 8 rolls of wallpaper and 1 packet of paste.

Emulsion paint
Peter needs enough emulsion paint for 2 coats on the living room ceiling, the kitchen ceiling, and the kitchen walls.

Living room ceiling area	$4.5 \times 3.2 = 14.4$	m²
Kitchen ceiling area	$3.4 \times 2.1 = 7.14$	m²
Kitchen walls area	$2 \times 3.4 \times 2.6 + 2 \times 2.1 \times 2.6 = 28.6$	m²
Total area	$14.4 + 7.14 + 28.6 = 50.14$	m²
For 2 coats, total area	$2 \times 50.14 = 100.28$	m²

1 litre of emulsion covers 11 m² so
10 litres would cover $10 \times 11 = 110$ m².
10 litres would be needed for Peter's job.

Gloss paint
Estimate of woodwork area is 7 m². For 2 coats, total area = 14 m².
1 litre covers 15 m² so 1 litre will be needed.

Estimate
Now that you know what materials Peter needs to buy, you can work out the total cost using the prices on p.128.

8 rolls of wallpaper at £4.35	= £34.80
1 packet of paste	= £ 1.69
4 2½ litre tins of emulsion at £7.25	= £30.00
1 litre tin of gloss paint	= £ 3.95
Total estimate	= £70.44

Decorating investigation

You have just moved into a new house and your bedroom needs completely redecorating. A plan of your room is shown below. You have a maximum of £200 to spend on the decoration and floor covering.

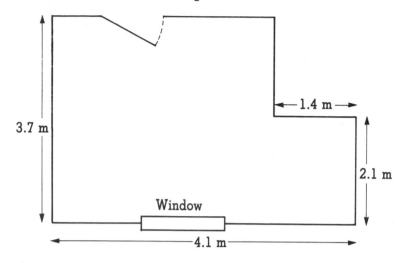

The height of the room is 2.8 m.
The door measures 1.2 m wide by 2.1 m high.
The skirting board is 12 cm high and runs right round the room.
The inside measurements of the window are 1.4 m wide by 1.8 m high. The frame is made from wood 15 cm wide.

The wooden surfaces will all need 2 coats of undercoat before they can be glossed.

Use this information to help you to decide what you will buy for £200.

Wallpaper
Woodchip £0.55 per roll (needs 2 coats of emulsion paint)
Patterned £2.55, £2.95, £3.25 or £3.65 per roll
Paste   £1.69 for up to 8 rolls of paper

Paint
Gloss   1 litre covers 15 m²
           £2.29 for ½ litre tin
           £3.95 for 1 litre tin
           £6.29 for 2½ litre tin
Emulsion   1 litre covers 11 m²
           £3.59 for 1 litre tin
           £7.25 for 2½ litre tin
Undercoat 1 litre covers 17 m²
           £3.95 for 1 litre tin
           £7.99 for 2½ litre tin

Floor covering
Cork tiles   £3.89 for pack of 9, 30 cm square tiles
Adhesive   £8.50 will stick up to 36 tiles
Carpet tiles £12.95 for pack of 9, 30 cm square tiles
Vinyl   £5.25 per linear foot (2 m wide)
              £6.99 per linear foot (3 m wide)
Carpet   £7.43 per square metre (3 m wide)

Newspaper investigation

When you buy a newspaper how much news do you actually get?
Get a copy of a daily paper. Look at it carefullly. What area of the paper would you say is taken up by adverts? 10%?, 20%?, 50%?, more?

Work with a partner. Calculate the area of 1 page, multiply by the number of pages. This gives the total area of newspaper. Share the paper between yourselves and calculate the total area of adverts.

Do the calculation $\dfrac{\text{Area of adverts}}{\text{Area of paper}} \times \dfrac{100}{1}$

Your answer is the percentage of the newspaper taken up by adverts. Does this figure surprise you? Did you expect it to be bigger or smaller?

Try other national and local papers. Do the same calculation for each, making a guess before you start. Compare these with your first paper. Can you think of reasons why different papers have different percentages of adverts?

Tiling investigation

Carpet tiles are made in the shape of a square of side 30 cm. They are made in a variety of colours. They are sold in packs containing 9 tiles all of the same colour. They cannot be bought singly.

The floor in the diagram is to be covered in carpet tiles, using two colours. Design different styles of laying out the tiles. For each one state the number of packs of each colour to be bought and the number of spare tiles. What is the most efficient way to organise your design?

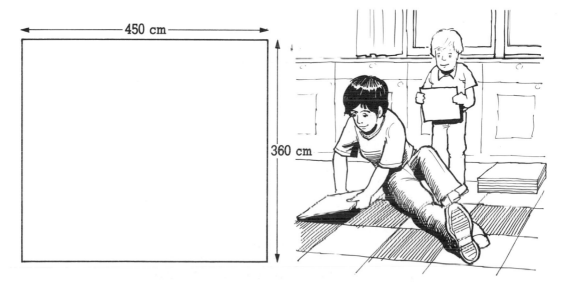

Reflection investigation

The coordinate diagram below shows 8 points labelled A to H. Reflect each point in turn in the x-axis to find its image. Draw the image of each point on the graph. Point A is shown as an example. The image of A is A′. Copy and complete the table below.

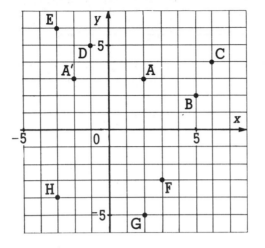

Point	Image after reflection in the x-axis
A (2,3)	A′ (−2,3)
B (5,2)	
C (6,4)	
D (−1,5)	
E (−3,6)	
F (3,−3)	
G (2,−5)	
H (−3,−4)	

What is the rule linking each point and its image when reflected in the x-axis?

Repeat the investigation to find the rule linking a point and its image after reflection in the y-axis.

The diagram opposite shows the line y=x. All the points on this line have the same x and y coordinates.

What rule links each point and its image after reflection in the line y = x?

Translation investigation

On squared paper plot the points A(5,3), B(2,1) and C(3,5). Join up the points to form a triangle. Move the triangle 3 squares to the right and 2 squares up. Copy and complete this table. What rule links each point and its image under the translation '3 right and 2 up'?

Point	Image
A(5,3)	(8,5)
B(2,1)	(,)
C(3,5)	(,)

Starting again with triangle ABC, move it 1 square right and 4 squares up. Copy and complete this table. What rule links each point with its image this time?

Point	Image
A(5,3)	(6,7)
B(2,1)	(,)
C(3,5)	(,)

What is the rule if the triangle is moved 7 squares to the left and 4 squares down?

Tessellations investigation

A regular tessellation is made by repeating the same shape over and over again so that no spaces are left uncovered. The Dutch artist M C Escher is famous for his paintings using tessellations. In the illustration entitled *Horseman*, each horseman fits exactly into its neighbours with no gaps.

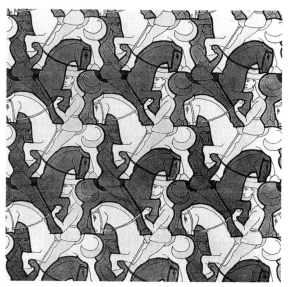

Here is a method for creating Escher-type tessellations.

Start with a simple shape which tessellates. Cut a piece out from one side and add it on to the opposite side. Repeat the process as many times as you need. With a little imagination you can produce tessellations like the one shown below. Try to create some interesting pictures yourself using this method.

Tessellation of squares

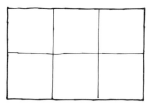

9 Maths abroad

Do you remember? ↩ In this unit it is assumed that you know how to:

1   calculate percentages

2   multiply and divide numbers with decimal fractions

3   express answers to calculations to a certain degree of accuracy.

Exercise 0 will help you check. Can you answer all the questions?

Exercise 0

Percentages
Calculate

1 10% of 153	2 10% of 257	3 10% of 84
4 5% of £165	5 25% of £84	6 20% of £1500
7 15% of £1250	8 20% of £850	9 5% of £60
10 12½% of £88	11 4½% of £156	12 6½% of £1000

Multiplying and dividing decimals
Calculate

13 150×11.2	14 250×5.6	15 340×2.2
16 180×6.5	17 400×7.8	18 550×9.2
19 240×126.57	20 380×214.11	21 265×12.657
22 $116 \div 6.4$	23 $284 \div 14.2$	24 $318 \div 2.4$
25 $44.8 \div 11.2$	26 $232 \div 3.2$	27 $125 \div 6.25$

Degrees of accuracy
Give the answer to these money calculations correct to the nearest 1p.

28 £325 ÷ 45	29 £670 ÷ 2.56	30 £34.50 ÷ 9
31 £45.50 × 10.72	32 £25.20 × 16.74	33 £80.45 × 6.67
34 £16.50 ÷ 7	35 £125 ÷ 9	36 £2.71 × 5.4

37   Three friends go to a restaurant for a meal. The bill comes to £38.50 which they agree to split equally. What is the minimum each person must pay?

38   Six people win a total of £10 000 on the football pools. The money is to be shared equally. How much will each person get?

Holiday money

Denzil and Harry have decided to go abroad for a holiday. They have heard that Majorca is worth visiting and Denzil has been to a travel agent and booked a 2 week holiday from 9-23 August. The last thing they have got to arrange is their foreign currency. To be safe they have decided to take £75 each in pesetas and the rest in travellers cheques.

Three weeks before they are due to leave Denzil goes to the bank to get his pesetas. Here is the notice board in the bank showing the exchange rates. Denzil and Harry will be buying spanish currency.

	WE SELL	WE BUY
Austria (sch)	19.15	20.80
Belgium (fr)	58.10	60.80
Denmark (kr)	10.53	10.88
France (fr)	9.23	9.63
Greece (dr)	184.00	197.00
Holland (gldr)	3.16	3.34
Italy (lire)	1925.00	2025.00
Norway (kr)	10.20	10.60
Spain (pes)	184.00	194.00
Sweden (kr)	9.56	9.96
West Germany (Dm)	2.80	2.97

Harry left it until one week before the holiday to get his pesetas. When he went to the bank the exchange rates had changed.

	WE SELL	WE BUY
Austria (sch)	19.50	21.10
Belgium (fr)	59.35	62.05
Denmark (kr)	10.76	11.11
France (fr)	9.35	9.74
Greece (dr)	184.00	196.00
Holland (gldr)	3.20	3.38
Italy (lire)	1960.00	2060.00
Norway (kr)	10.30	10.80
Spain (pes)	187.00	197.00
Sweden (kr)	9.69	10.08
West Germany (Dm)	2.83	3.01

Who got the better deal, and by how much?

This unit will explain about holidays abroad and help you solve problems involving foreign currency and exchange rates.

Holiday prices

Information ⇨ Below is a table which gives the **basic price** of skiing holidays to Bormio in Italy, based on season 1985/86. The price depends on

(a) when you are leaving for your holiday

(b) the hotel you choose to stay in

(c) the length of your holiday (7 or 14 nights).

In the table the letters BB stand for bed and breakfast, letters HB stand for half board. Half board includes bed, breakfast and evening meal.

Hotel	Gufo		San Lorenzo		Funivia		Children's Discount
Board Basis	BB		HB		HB		
Holiday No.	BM47		BM45		BM46		
Nights in Resort	7	14	7	14	7	14	
Dec 22	251	373	261	399	294	477	10%
Dec 29	268	344	281	382	311	419	15%
Jan 5, 12	181	279	191	289	214	315	Free
Jan 19, 26	193	308	199	318	219	326	Free
Feb 2	211	319	218	329	232	345	Free
Feb 9	222	331	230	341	239	368	15%
Feb 16	252	342	259	353	265	379	10%
Feb 23	221	328	227	337	235	364	20%
March 2	210	309	215	319	222	347	30%
March 9, 16	199	287	201	294	207	322	Free
March 23, 30	214	312	217	321	224	338	10%
April 6	195	—	197	—	195	—	Free

Example

Individual cost

How much would it cost Lorraine to go on a skiing holiday to Bormio if she were to leave on 16 February and stay in the Hotel San Lorenzo for 14 nights?

There are 3 pieces of information given: (a) 16 February, (b) Hotel San Lorenzo, (c) 14 nights

Look down the column headed San Lorenzo, 14, and along the row starting with Feb 16. The row and column meet at 353. The basic price of Lorraine's holiday will be £353.00.

Example

With discount

Mr and Mrs Scott and their 2 children, Patricia and Alison, decide to go to Bormio for 14 nights at the same time as Lorraine, staying in Hotel San Lorenzo. What is the cost for (**a**) each child (**b**) the whole family?

(**a**)   The basic price is £353.00
Children get a 10% discount.
10% of £353 = £35.30
Cost per child is £353.00−£35.30 = £317.70

(**b**)   2 adults   at £353.00 = £ 706.00
2 children at £317.70 = £ 635.40

Total cost   = £1341.40

Exercise 1

Use the table on the previous page to find the cost of the holidays below.

	Departure date	Nights in resort	Hotel	Basic price
1	23 February	7	Gufo	
2	29 December	14	Funivia	
3	2 March	14	Funivia	
4	2 February	7	San Lorenzo	
5	6 April	7	Gufo	
6	22 December	14	San Lorenzo	
7	16 March	7	Gufo	
8	26 January	14	Funivia	
9	30 March	7	San Lorenzo	
10	6 April	14	Gufo	

11   Explain your answer to question **10**.

12   Which of the 3 hotels is cheapest? Give a reason for your answer.

13   Of the 2 hotels which offer half board which is the cheaper?

14   Which are the cheapest departure dates for all the hotels?

15   Which is the most expensive departure date for all the hotels?

16   Why do you think the prices vary from week to week?

17   Find the cost of a skiing holiday in Bormio for Mr and Mrs Thomson. They are leaving on 2 March and are staying in the Hotel Funivia for 7 nights.

18   The Peters family have decided to have a skiing holiday in Bormio, staying in the Hotel Gufo for 14 nights, departing on 9 February. What will the holiday cost Mr Peters for himself, his wife and 3 children?

19   Mr and Mrs Bashir and their daughter Tabassum are going to Bormio for a 2 week skiing holiday. They are leaving on 19 January and staying in the Hotel San Lorenzo. What will the holiday cost?

20   Angela Graham is taking her 2 children skiing to Bormio. They are staying in the Hotel San Lorenzo for 7 nights and are leaving on 23 February. How much will the holiday cost?

Hidden extras

Information

For most skiing holidays there are a few extra costs to be added onto the basic price. Some are compulsory and some optional. Insurance, airport tax and a flight supplement if you are leaving from a different airport, are compulsory. For skiing holidays, lift passes, ski hire and ski instruction are optional. Here is the information from a brochure on Bormio.

Insurance   Up to 10 days £15.45   Up to 18 days £18.80

Lift pass

	Adults		Children	
Days	6	13	6	13
High season	£46.00	£73.00	£41.50	£62.50
Low season	£34.00	£60.00	£30.00	£50.00

(High season dates 21 December to 3 January and 1 February to 4 April)

Ski equipment and ski school

	Adults		Children	
Days	6	13	6	13
Boots	£ 5.00	£ 9.00	£ 5.00	£ 9.00
Skis/sticks	£10.50	£20.00	£10.50	£20.00
Ski school	£24.00		£24	

Example

Calculating the extra cost

Mr and Mrs Scott have their own equipment but need to hire skis, sticks and boots for their 2 children. They must also book lift passes for all the family. Insurance is compulsory for everyone. How much will this add to the holiday price? (Remember that their departure date is 16 February so they will need High Season lift passes.)

2 Adult 13 day lift passes at £73.00 each	= £146.00
2 Child lift passes at £62.50 each	= £135.00
2 Children's boot hire at £9.00 each	= £ 18.00
2 Children's ski/stick hire at £20.00 each	= £ 40.00
4 Insurances at £18.80 each	= £ 75.20
Total to be added to the original price	= £414.20

Total bill for the holiday would be (see p.156) £1342.40 + £414.20 = £1756.60

Exercise 2

Calculate the additional payments that are due for each of the skiing holidays to Bormio described in questions **1-5**.

1   Mr Graham is going for a 1 week holiday, leaving on 2 February and staying in the Hotel Gufo. He has his own equipment and does not want ski instruction.

2 Billy has never skied before and has booked a 14 day holiday to Bormio leaving on 12 January, staying in the Hotel San Lorenzo. He has decided to have ski lessons for the first week only. He will need to hire skis, sticks and boots and book a lift

3 Brenda is an experienced skier who has booked a 2 week holiday to Bormio, departing on 9 March, staying in the Hotel Gufo. She has her own equipment and has decided to have ski lessons for the first week only. She will also need a lift pass.

4 Mr and Mrs Marshall are going skiing to Bormio for 1 week leaving on 6 April. They are both beginners and need to hire skis, sticks, and boots, and also have ski instruction and lift passes. They are staying in the Hotel Funivia.

5 Mr and Mrs Balfour and their 4 children are going to Bormio for 7 days skiing leaving on 19 January. Mr and Mrs Balfour are expert skiers and only require lift passes. The children all need to hire equipment, and also have ski instruction and lift passes. They have decided to stay in the Hotel San Lorenzo.

6 For each of the holidays in questions **1-5** above
   (a) use the table on page 155 to find the basic price,
   (b) calculate the total holiday price once all the additions have been included.

Bar charts

Information ☞ Brochures advertising summer holidays give a lot of information about each resort. Some of the information is in tables and some in the form of bar charts. Often the information is about the temperature at various times in the year.

Example

Interpreting a bar chart
The bar chart shows the average daily temperature in the Costa Brava compared with London.

(a) Which is the hottest month in the Costa Brava?
(b) Estimate the average daily temperature in July in the Costa Brava.
(c) How much hotter is it in the Costa Brava in September compared with London?

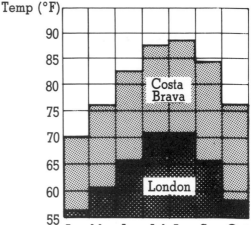

(a) The hottest month is August.
(b) Each square on the vertical axis represents 5 degrees. We have to estimate the answer to the nearest degree. The top of the column lies between 85 and 90. Estimate 88 degrees.
(c) Costa Brava temperature in September   84 degrees
    London temperature in September   66 degrees

    Difference in temperature in September   18 degrees

Example

Drawing a bar chart
The table below gives the average daily hours of sunshine per month in the Costa Brava and London. Show this information in a bar chart.

Month	Apr	May	Jun	Jul	Aug	Sep	Oct
Costa Brava	9	10	11	12	11	9	7
London	6	7	8	7	6	5	4

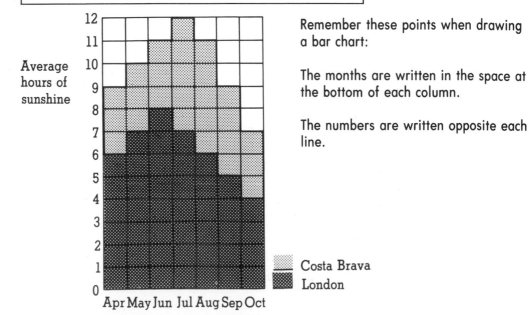

Remember these points when drawing a bar chart:

The months are written in the space at the bottom of each column.

The numbers are written opposite each line.

Exercise 3   The bar charts show the average daily temperature for different resorts. For each resort find

(a) the hottest month
(b) the average daily temperature that month
(c) the month with the largest temperature difference from London
(d) the temperature difference that month.

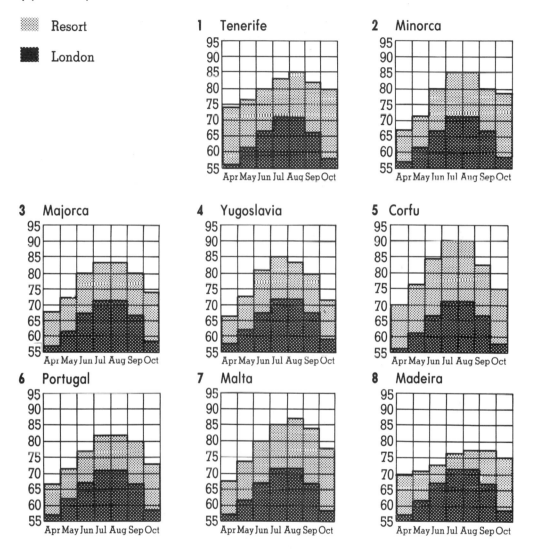

The table below gives the average daily hours of sunshine for various resorts. Construct bar charts to display the information. Also show the London equivalent on each bar chart.

		Apr	May	Jun	Jul	Aug	Sep	Oct
	London	6	7	8	7	6	5	4
9	Tenerife	8	9	10	11	10	8	6
10	Minorca	7	10	10	11	11	8	4
11	Yugoslavia	6	8	10	12	11	9	7
12	Corfu	7	9	11	13	12	9	6
13	Greek Islands	8	10	14	14	13	11	8
14	Malta	8	10	11	11	11	9	7
15	Madeira	8	7	7	8	8	7	7

Averages

Information ⇨ So far the tables and bar charts you have looked at have all shown **average** figures for sunshine or temperature. For example the average daily temperature in June on the Costa Brava is 83 degrees. It does not mean that the temperature will rise to 83 degrees everyday in June. Some days will be cooler and some hotter. You may sometimes see the word **mean** used instead of average.

Example

On an expedition to Morocco, Stewart Robertson recorded the temperature (°F) at the same time every day. Here are his results

Day 1	Day 2	Day 3	Day 4	Day 5	Day 6	Day 7	Day 8	Day 9	Day 10	Day 11	Day 12
85	89	88	92	94	90	86	86	82	85	84	78

What was the average temperature Stewart recorded?

To find the average, add up all the temperatures and divide by 12, because there are 12 recorded temperatures.

$$(85+89+88+92+95+93+88+86+82+85+84+78) \div 12$$

$$= 1039 \div 12 = 86.58 \text{ (rounded off to the nearest degree is 87)}$$
Average temperature was 87°F

Exercise 4

Bryan Ryalls is an amateur meteorologist who lives near Edinburgh. Here are his records for the first 14 days in June 1985.

1 Temperature (°C)

Jun 1	Jun 2	Jun 3	Jun 4	Jun 5	Jun 6	Jun 7
21.9	20.8	23.1	17.5	11.4	11.8	11.8

Jun 8	Jun 9	Jun 10	Jun 11	Jun 12	Jun 13	Jun 14
13.0	15.1	16.0	15.5	13.3	11.2	13.0

2 Rain fall (mm)

Jun 1	Jun 2	Jun 3	Jun 4	Jun 5	Jun 6	Jun 7
0	0	0	0	0	0.25	2.29

Jun 8	Jun 9	Jun 10	Jun 11	Jun 12	Jun 13	Jun 14
0	3.30	1.78	8.13	9.40	3.30	5.08

3 Sunshine (hrs)

Jun 1	Jun 2	Jun 3	Jun 4	Jun 5	Jun 6	Jun 7
13.8	15.2	14.9	13.2	1.0	0.8	2.6

Jun 8	Jun 9	Jun 10	Jun 11	Jun 12	Jun 13	Jun 14
4.4	5.0	9.3	0.1	1.0	1.4	5.0

(a) For each record find the average for the period.
(b) Draw a bar chart of each of these records and mark the average for the period with a dotted line.

4 In an experiment 15 people measured how long they could hold their breath. Their results are given below in seconds. Calculate the average.

24 31 19 24 27 18 30 27 25 28 28 26 19 21 34

Do this in class and see how this average compares with your class average.

5 The first 9 scores in a darts match are shown below. Find the mean score.

35 26 46 27 60 35 71 25 31

6 The ages of the players in a local football team are given below. Calculate the mean age.

19 23 25 24 19 25 31 27 28 30 33

The 2 oldest players leave and are replaced by 2 players aged 18 and 26. Calculate the new mean.

7 In a 5-a-side football competition the average age of a team must not exceed 16. Below are the ages of 2 groups of 10 players who each hope to make up 2 teams. How would you arrange the ages?

(a) 14 16 14 17 15 18 16 15 17 18
(b) 14 15 16 17 15 16 14 16 18 14
(c) Here are the ages of 1 team. Will they be allowed to take part in the competition? 15 17 16 17 16

8 In a skating competition, Tanya scores an average mark of 5.3. Seven of her eight marks were 4.9 5.6 4.8 4.9 5.8 5.5 5.3. What was her eighth mark?

9 The average height of a basketball team of 12 players is 2.02 m. Joe, who is 1.92 m tall, leaves the team. When Floyd joins as his replacement the average height of the team goes up to 2.03 m. How tall is Floyd?

10 The average weight of the school rugby team is 70 kg. If the star player, weighing 65 kg, is replaced by someone weighing 80 kg, what is the new average?

Surcharges

Information ➦ Holiday companies have to set their prices many months before the advertised holiday dates. Changes in the value of foreign currency, or in the price of fuel can affect the price which the company has to charge. To account for this some travel companies add on a **surcharge**. This is a percentage increase on the basic holiday price.

Example
Calculating a surcharge
The travel company which the Scott family is using for their skiing holiday in Bormio is adding a 4% surcharge to the basic price. What will their final bill come to?

In the example on page 157 the Scott's bill was calculated. It was made up of 2 parts — the basic price and extras.

Basic Price	=	£1342.40
Surcharge, 4% of £1342.40	=	£ 53.70
Extras	=	£ 414.20
Final price	=	£1810.30

Exercise 5

1   Find the final cost of a holiday to Spain if a 5% surcharge is added to the basic price of £264.

2   The basic price of a holiday to Portugal is £326. Find the final price if a 6% surcharge is added.

3   A wine tasting holiday in France costs £168. How much will the holiday cost if a 4½ % surcharge is added?

4   The basic price of a holiday in Greece is £245. Excursions and outings cost an extra £85. Find the total price if a 7% surcharge is added to the basic price.

5   A holiday in Malta costs £306. Water-skiing and windsurfing cost an extra £65. Find the total cost if a surcharge of 6½ % is added to the basic price.

6   A 2 week holiday in Ibiza costs £252 per person. Car hire is extra and costs £65 per week. The company add on a surcharge of 3% of the basic price. Find the total cost of the holiday including 2 weeks car hire and insurance costing £16.50 each.

7   The basic cost of a 14 day holiday in Tunisia is £271. Children get a reduction of 30%. Insurance costs £15.40 each, and a 5% surcharge is to be added to the basic price. Find the total cost of the holiday for Mr and Mrs Roberts and their son David.

Money abroad

Information ⇦ Different countries use different **currencies**. That is, different notes and coins. If you intend to go abroad on holiday you should know which currencies are used in the countries you are visiting.

Example **Countries and currency**
In France the **unit of currency** is the Franc.
In Spain the unit of currency is the Peseta.

Exercise 6 1 Here is a map of Europe. Copy or trace the map and mark in the countries. (You may need an atlas to help you).

2 Copy and complete the table below.

Country	Capital	Currency
Portugal	Lisbon	Escudos
Spain		
France		

Investigation The dollar is used as a unit of currency in many different countries. List as many of these countries as possible.

Exchange rates

Information ⇆ If you go abroad you must use the currency of the country you visit. The safest way of changing money is to get **traveller's cheques** in this country which can be exchanged for cash in a bank in the country you visit. There their value will be calculated using the **exchange rate**. This rate tells you how much foreign currency you can buy for £1.

The exchange rate varies from time to time. Below is a table showing the rates for buying the currency of popular holiday countries, correct at the time of writing.

Austria (sch)	19.50	Italy (lire)	1960.00
Belgium (fr)	59.35	Norway (kr)	10.39
Canada ($)	1.98	Portugal (esc)	203.00
Denmark (kr)	10.76	Spain (pes)	187.00
France (fr)	9.35	Sweden (kr)	9.69
Greece (dr)	184.00	Switzerland (fr)	2.28
Holland (gldr)	3.20	United States ($)	1.43
Ireland (punts)	1.05	West Germany (Dm)	2.83

Examples

Buying foreign currency

1 Lorraine, who is going to Bormio in Italy for a skiing holiday, takes £500 with her. How many lire would she get?

$$£1 = 1960 \text{ lire}$$
$$£500 = 1960 \times 500 \text{ lire}$$
$$= 980\,000 \text{ lire}$$

2 Paul is going to stay with a family in France for 2 weeks. How many Francs will he get for £50?

$$£1 = 9.35 \text{ Francs}$$
$$£50 = 9.35 \times 50 \text{ Francs}$$
$$= 467.5 \text{ Francs}$$

Exercise 7

Use the exchange rate above to answer questions 1-5 by completing the table below.

	Country	Amount to exchange (£)	Foreign currency
	Belgium	100	59.35 × 100 = 5935 fr
1	Holland	250	
2	Portugal	300	
3	Switzerland	425	
4	Italy	80	
5	France	140	

6 Mr Sandhu arrives in Norway with £250 worth of traveller's cheques. How many Kroner would he get for this?

7 The Robinson family go to Switzerland for a holiday. While there they cash cheques valuing £60, £130, £50, £80, and £75. How many Swiss Francs would they be given in total?

8 The Smiths go on a mini-tour of Europe during which they cash cheques for £100 in Denmark, £50 in Norway and £210 in Sweden. How much of each currency did they have to spend?

Coming home

Information ☞ When you come back from abroad you may have some foreign money left over which you will want to sell back for **Sterling** (British money). Banks will exchange foreign notes but not coins. Remember you *multiply* to change from Sterling to foreign currency and *divide* to change from foreign currency to Sterling.

The exchange rates for selling foreign money are different to the rates for buying. You usually lose a bit of money if you come back with a lot of extra foreign money. These rates are also shown on boards in banks. An example of rates is shown below.

Austria (sch)	21.10		Italy (lire)	2060.00
Belgium (fr)	62.05		Norway (kr)	10.80
Canada ($)	2.05		Portugal (esc)	217.00
Denmark (kr)	11.11		Spain (pes)	197.00
France (fr)	9.74		Sweden (kr)	10.08
Greece (dr)	196		Switzerland (fr)	2.46
Holland (gldr)	3.38		United States ($)	1.49
Ireland (punts)	1.03		West Germany (Dm)	3.01

Examples

Selling foreign currency

1 Lorraine has 15 000 lire left after her holiday in Italy. How many pounds will she get for this?

£1 = 2060 lire
15 000 ÷ 2060 = £7.28

2 Paul has 50 Francs left after his stay in France. How much will he get in Sterling?

£1 = 9.74 Francs
50 ÷ 9.74 = £5.13

Exercise 8

Use the exchange rates above to answer questions **1-6** by completing the table below.

	Country	Money left	British equivalent
1	Canada	74 dollars	
2	Belgium	165 francs	
3	Germany	46 D'marks	
4	Portugal	850 escudos	
5	Eire	23 punts	
6	Sweden	125 kroner	

7 Lorna bought a Swiss Cuckoo clock as a present. It cost 38 Swiss Francs. How much is this in Sterling?

8 Mr and Mrs Munro paid 85.5 D'marks for a meal in a German restaurant. How much is this in British money?

9 Lee Wang and his family went on holiday to France. They took with them £350 spending money.
(a) How many Francs is this?
(b) How much would they have left if they spent 3250 Francs on holiday?
(c) How much is this in Sterling?

10 Stan and Sheila Tait take a trip on concorde to New York. If they took £250 with them how many dollars would they get? If they used 320 dollars, how much British money would they get back?

Holiday money

Denzil: 3 weeks before holiday

	WE SELL	WE BUY
Austria (sch)	19.15	20.80
Belgium (fr)	58.10	60.80
Denmark (kr)	10.53	10.88
France (fr)	9.23	9.63
Greece (dr)	184.00	197.00
Holland (gldr)	3.16	3.34
Italy (lire)	1925.00	2025.00
Norway (kr)	10.20	10.60
Spain (pes)	184.00	194.00
Sweden (kr)	9.56	9.96
West Germany (Dm)	2.80	2.97

Harry: 1 week before holiday

	WE SELL	WE BUY
Austria (sch)	19.50	21.10
Belgium (fr)	59.35	62.05
Denmark (kr)	10.76	11.11
France (fr)	9.35	9.74
Greece (dr)	184.00	196.00
Holland (gldr)	3.20	3.38
Italy (lire)	1960.00	2060.00
Norway (kr)	10.30	10.80
Spain (pes)	187.00	197.00
Sweden (kr)	9.69	10.08
West Germany (Dm)	2.83	3.01

You can see from the notices that the bank was selling pesetas at the rate of 184 per £1 when Denzil changed his money and at the rate of 187 per £1 when Harry changed his money.

Denzil
£ 1 = 184 pesetas
£75 = 184 × 75 pesetas
    = 13 800 pesetas

Harry
£ 1 = 187 pesetas
£75 = 187 × 75 pesetas
    = 14 025 pesetas

Harry got 225 more pesetas than Denzil.

Investigation

Exchange rates
Each day newspapers give the exchange rates for the main world currencies. Note the exchange rate for one currency every day for a month. Draw a graph of your results to show how the rate has varied. What is the trend of your graph?

How does your chosen currency compare with others? Look at the results from other students in your class.

Investigations **Planning a package holiday**
Get a holiday brochure and plan a foreign package holiday for yourself and 2 of your friends.

Each of you should answer this list of questions.

1 How much can you afford to pay for the holiday?

2 What time of year you would like to take a holiday?

3 What country or countries do you want to visit?

4 Do you want to stay in a hotel or apartment?

5 If you choose a hotel do you want full or half-board?

6 What length of holiday should you plan (7 or 14 nights)?

7 Would you like lots of nightlife (Bars, Discos etc)

8 Would you like sports to be available?

Compare your answers and come to some compromise — remember, you can't always get what you want!

Once you have agreed on the holiday you should complete the booking form in the brochure.

You should also get and complete:
   a passport form,
   an insurance form if required,
   form E111 from the DHSS.

You should also arrange for foreign currency.
Decide how much you want in travellers cheques.
Decide how much you want in currency.

After completing all the information you will probably need a holiday!

10 Sport

Do you remember? ⇨ In this unit it is assumed that you know how to:

1 calculate the average from a given list of numbers
2 draw a bar chart from a given list of data
3 use percentages.

Exercise 0 will help you check. Can you answer all the questions?

Exercise 0 **Calculating the average from a list of numbers**

Calculate the mean, correct to 1 decimal place, for each of these lists of data.

1 15, 23, 45, 27, 15

2 72, 48, 23, 46, 12, 72, 46

3 65, 24, 67, 28, 43, 37, 25, 36

4 11, 34, 8, 13, 35

5 11.5, 23.5, 15.2, 16.8

6 7.3, 2.5, 3.8, 2.8, 2.5, 3.9

Drawing bar charts from a list of data

7 The goals scored in weekly matches by a local football team are shown below.

Goals	0	1	2	3	4	5	6
Frequency	3	4	2	5	2	1	0

Show the information on a bar chart.
How many football matches have been played?

8 The daily sales of crisps from a school shop are shown below.

Day	Mon	Tue	Wed	Thu	Fri
Sales	27	29	22	36	18

Show the information on a bar chart.

Using percentages in solving problems

9 When he started his diet Sam weighed 120 kg. After 4 weeks his weight had fallen by 5%. How much did he weigh after 4 weeks?

10 Before going on holiday Grant weighed 76 kg. When he came back he weighed 79.5 kg. Express his gain in weight as a percentage of his weight before he left for holiday.

New car competition

WIN A NEW CAR

Listed below are 10 important features of the new FIAT PANDA range

a Good fuel economy
b Dual circuit brakes
c Fold down rear seat
d Five-speed gearbox
e Single overhead camshaft
f Good all round vision
g Rear screen wiper and washer
h Tinted glass
i Sun roof
j Front-wheel drive

Put these features in order of importance and, if your order agrees with our panel of judges, you will win a new FIAT PANDA.

Dermot sees this competition form in a local car showroom.

Dermot got an entry form and decided to have a shot at winning the car.

In which order would you place the 10 features?

What chance does Dermot stand of winning the car?

This unit will help you decide.

Knock-out competitions

Information ➪   John and Sebrina are organising 2 competitions at their Youth Club.
John is organising a pool **knock-out** competition and Sebrina is in charge of the table tennis **league**.

Example   **Knock-out competition**
16 people enter John's pool competition. They are Peter, Waseem, Leslie, Assad, Harry, Keith, Tom, Steve, Mumtaz, Terry, Linda, Fiona, John, Gary, Fiaz and Alan.

John draws each name out of a hat and writes them down in a column showing the first round games. The winners name goes forward into the next round and so on. This is how the competition worked out.

First round	Quarter final	Semi final	Final
Peter			
Terry	Terry		
Fiona		Terry	
John	Fiona		
Alan			Mumtaz
Mumtaz	Mumtaz		
Steve		Mumtaz	
Keith	Steve		
Harry			Linda
Assad	Harry		
Waseem		Harry	
Gary	Gary		
Tom			Linda
Fiaz	Fiaz		
Linda		Linda	
Leslie	Linda		

This shows that Linda won the competition having played and won 4 games.

Exercise 1   For questions **1-7** look at the order of the knock-out competition on the previous page and decide

1 who was runner up.

2 which players got through to the semi-final.

3 how many games Fiaz won.

4 which players were knocked out in the first round.

5 which players were knocked out in the quarter-final.

6 which players played most games in the competition.

7 how many games were played altogether.

8 You have seen that a knock-out competition can be organised for 16 people. Can one be simply organised for 24 people? Give a reason for your answer.

9 What happens if 30 people enter a knock-out competition?

10 The smallest number of people that can be used for a knock-out competition is 2. List the next 4 numbers of people that can be used for knock-out competitions, assuming that everyone plays in the first round.

11 How many people would get straight to the second round if 20 people entered a knock-out competition?

12 Copy and complete the table.

Number of people (n)	2	4	8	16	32	64
Number of rounds (r)	1		3			
Number of games (g)	1			15		

What is the relationship between n the number of people and g the number of games?

13 Use the results from question **12** to predict the number of rounds and games in a knock-out competition in which 132 people enter.

14 How many games would be needed in a knock-out competition if 22 players entered? How would it be organised?

15 20 people enter the 200 metres in the school sports. How would you organise the heats and the final. Your school running track has 8 lanes.

Investigation   Organise a knock-out competition for your class. Put blank pieces of paper into the hat if necessary to make the names up to a correct starting number. Draw names out of a hat for the order of play. If someone draws out a blank piece then they can go straight through to the next round. Choose a simple game, for example noughts and crosses on a 4×4 grid.

League competitions

Information ⟿ The table-tennis league Sebrina is running is different to John's knock-out competition. In the **league** competition everyone plays everyone else once. Each match consists of 4 games. A player gets 2 points for winning a match, 1 point for a draw, and nothing for losing. The winner of the league is the person with the most points at the end.

For each player there is also a score **for** (the total number of games won) and **against** (the total number of games lost). So if 2 players have the same number of points, the one with the best score (**for** total – **against** total) is placed higher.

Example **Preparing a league table**

Here are the results for the first 3 rounds.

Round 1			Round 2			Round 3		
Sajid	3	Jim	1	**Sajid**	2	Bob	2	**Sajid**	1	Keith	3
Darren	2	Bob	2	Darren	4	Keith	0	Darren	2	Isla	2
Ruth	1	Keith	3	Ruth	2	Isla	2	Ruth	4	Jim	0
Karen	4	Isla	0	Karen	3	Jim	1	Karen	2	Bob	2

Here is the league table after 3 rounds.

	Played	Won	Drawn	Lost	For	Against	Points
Karen	3	2	1	0	9	3	5
Darren	3	1	2	0	8	4	4
Keith	3	2	0	1	6	6	4
Ruth	3	1	1	1	7	5	3
Sajid	**3**	**1**	**1**	**1**	**6**	**6**	**3**
Bob	3	0	3	0	6	6	3
Isla	3	0	2	1	4	8	2
Jim	3	0	0	3	2	10	0

Sajid has played 3 games, won 1, drawn 1 and lost 1. He has scored a total of 6 games for, 6 games against and has 3 points.

Exercise 2 Answer questions **1–5** using the information in the league table above.

1 Why is Darren above Keith in the league when they both have 4 points?

2 Why is Sajid above Bob in the league when they both have 3 points and the same games for and against?

3 Here are the results for the next 2 rounds of the competition.

Round 4			Round 5		
Sajid	2	Isla	2	Sajid	0	Darren	4
Darren	3	Jim	1	Isla	2	Jim	2
Ruth	2	Bob	2	Ruth	1	Karen	3
Karen	3	Keith	1	Bob	2	Keith	2

Redraw the league table to show the position after the fifth round.

4 Here are the remaining games that have to be played

Which players could win the league?

Round 6			Round 7		
Sajid	v	Ruth	Sajid	v	Karen
Isla	v	Bob	Darren	v	Ruth
Darren	v	Karen	Isla	v	Keith
Jim	v	Keith	Jim	v	Bob

5 Here are the results for the last 2 rounds.

Redraw the league table to show the final positions of the players.

Round 6				Round 7			
Sajid	2	Ruth	2	Sajid	1	Karen	3
Isla	4	Bob	0	Darren	4	Ruth	0
Darren	2	Karen	2	Isla	1	Keith	3
Jim	1	Keith	3	Jim	2	Bob	2

6 Here is an incomplete league table for the home international youth football competition. Two points are given for a win and 1 for a draw. Copy and complete the table.

Country	Played	Won	Drawn	Lost	For	Against	Points
Scotland	3	2	1		8	5	
England	3	1			8	7	4
Wales	3	0			4	6	
Ireland	3				2	4	1

7 The table shows one way of arranging games for a league of 4 netball teams. Each team should have a game to play each week of the competition. Each match is shown twice. For example in week 2 Team 3 plays Team 1 so Team 1 must also be shown as playing Team 3. Copy and complete the table.

Team	Week 1	Week 2	Week 3
1		3	4
2	1		3
3	4	1	2
4		2	

8 Copy and complete the table below which shows the order of matches for a league of 6 rugby teams.

Team	Week 1	Week 2	Week 3	Week 4	Week 5
1	2	3	4	5	6
2	1	4	5	6	
3	5		6	4	
4	6			3	
5		6			

9 Draw up a table to show the matches for a league of

(a) 8 hockey teams   **(b)** 10 volleyball teams

Olympic records

Information ➥ The Olympic Games are held every 4 years. Athletes compete in many sports and detailed records are kept of the results. The best ever recorded performance in each sport sets the **Olympic record** for that sport. Graphs are a good way of showing information.

The way a graph slopes tells you whether the thing being measured is **increasing or decreasing**. This is called the **trend** of the graph. Sometimes the trend of a graph can be used to predict what might happen in the future.

Example

The graph opposite shows how the winning throw for the Women's Javelin has changed over the years of Olympic competition.

Looking at the graph we can see that:

(a) The trend of the graph is upward.

(b) The only year in which the previous Olympic record was not beaten was 1968.

(c) The year in which the increase in the record was greatest was 1952.

(d) The Olympic Games are held every 4 years.

(e) There were no Olympic Games held in 1940 or 1944.

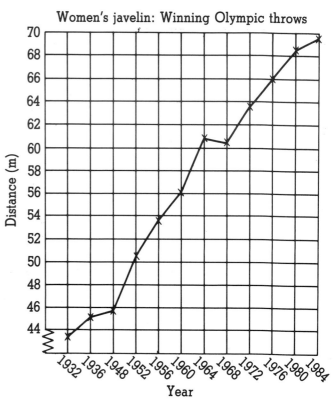

Example

The graph opposite shows how the winning time for the women's 4×100 m relay has changed over the years of Olympic competition.

From the graph you can see that:

(a) The trend of the graph is decreasing.

(b) The Olympic record was not beaten in 1948, 1960 or 1972.

(c) The year in which the greatest decrease in the record occurred was 1952.

Exercise 3

1 The graph below shows how the winning time for the Men's 200 metres has changed over the years of Olympic competition.

(a) What is the trend of the graph?

(b) In which years was the existing Olympic record not beaten?

(c) In which year was the decrease in the Olympic record greatest?

(d) What was the longest spell that the Olympic record went unbeaten?

(e) In which years were there no Olympic Games?

(f) Why do you think there were no Olympic Games in these years?

2 The graph below shows how the winning times for the Men's 1500 meters has changed over the years of Olympic competition.

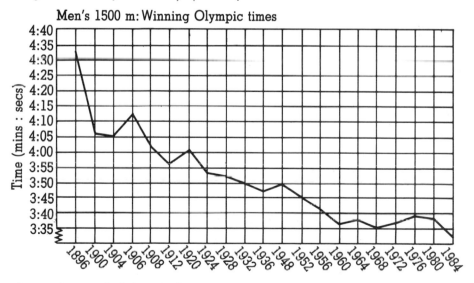

(a) What is the trend of the graph?

(b) In which year was the greatest time taken off the record?

(c) In which years was the previous winning time not beaten?

(d) In which years was the existing Olympic record not beaten?

(e) In which year did the record first fall below 4 minutes?

(f) Can you predict when the Olympic record will fall below 3 minutes 30 seconds?

3 The table below shows the total number of medals won by the top 12 countries in the 1984 Olympic Games. Show the information on a bar chart.

Country		Medals
Britain	GBR	37
Canada	CAN	44
China	ROC	32
France	FRA	27
Italy	ITA	32
Japan	JPN	32

Country		Medals
New Zealand	NZL	11
Romania	ROM	53
S Korea	KOR	19
USA	USA	174
W Germany	GER	59
Yugoslavia	YUG	18

Why do you think the USA won so many medals?

4 The table below shows the development of the Summer and Winter Olympic Games from 1896 to 1976. Show this information on separate bar charts. Draw 2 bar charts for each Olympics, Summer and Winter. Label them carefully.

Year	Summer Olympics		Winter Olympics	
	Countries Represented	Number of Sports	Countries Represented	Number of Sports
1886	13	9	—	—
1900	22	17	—	—
1904	12	14	—	—
1906	20	11	—	—
1908	22	21	—	—
1912	28	14	—	—
1920	29	22	—	—
1924	44	18	16	5
1928	46	15	25	6
1932	47	15	17	5
1936	49	20	28	6
1948	58	18	28	7
1952	69	17	22	6
1956	71	17	24	6
1960	83	17	27	5
1964	93	19	36	7
1968	112	18	37	7
1972	122	21	35	7
1976	88	21	36	7

5 The table below shows the medals won by the top 15 countries in the Winter Olympic Games up to 1976. Put the countries into alphabetical order and show the total number of medals won by each country on a bar chart.

Country		Gold	Silver	Bronze
USSR	URS	51	32	35
Norway	NOR	50	52	43
USA	USA	30	38	28
Sweden	SWE	25	23	26
Finland	FIN	24	34	23
Austria	AUT	22	31	27
W Germany	GER	22	19	17
Switzerland	SUI	15	17	16
E Germany	GDR	12	10	16
France	FRA	12	9	12
Canada	CAN	12	8	13
Italy	ITA	10	7	7
Netherlands	HOL	9	13	9
Great Britain	GBR	5	4	10
Czechoslovakia	TCH	2	5	6

6 The 1928 record for weightlifting was 327.5 kg. In 1972 a new record of 580 kg was set. What is the percentage increase in the weightlifting record from 1928?

7 Anna's best time for running the 1500 metres was 5 minutes 2.3 seconds. At a recent race meeting she won the 1500 metres in 4 minutes 58.9 seconds. By how much had she beaten her personal best time?

8 Here are the results for 4 different Olympic sports over the years of Olympic competition. Draw a graph of each event and answer the following questions.
(a) What is the trend of each graph?
(b) In which years was the existing Olympic record not broken?
(c) In which year was the change in the record greatest?
(d) What is the longest time that a record has stood unbeaten?

Men's 100 metres hurdles

Year	Winner		Time(s)
1896	T Burke	USA	12.0
1900	F Jarvis	USA	11.0
1904	A Hahn	USA	11.0
1906	A Hahn	USA	11.2
1908	R Walker	SAF	10.8
1912	R Craig	USA	10.8
1920	C Paddock	USA	10.8
1924	H Abrahams	GBR	10.6
1928	P Williams	CAN	10.8
1932	E Tolan	USA	10.3
1936	J Owens	USA	10.3
1948	H Dillard	USA	10.3
1952	L Remigino	USA	10.4
1956	R Morrow	USA	10.5
1960	A Hary	GER	10.2
1964	R Hayes	USA	10.0
1968	J Hines	USA	9.9
1972	V Borzov	URS	10.14
1976	H Crawford	TRI	10.06
1980	A Wells	GBR	10.25
1984	C Lewis	USA	9.99

Women's 200 metres

Winner		Time(s)
F Blankers-Koen	HOL	24.4
M Jackson	AUS	23.7
B Cuthbert	AUS	23.4
W Rudolph	USA	24.0
E McGuire	USA	23.0
I Kirszenstein	POL	22.5
R Stecher	GDR	22.4
B Eckert	GDR	22.37
B Wockel	GDR	22.03
V Brisco-Hooks	USA	21.81

Men's high jump

Year	Winner		Height (m)
1896	E Clark	USA	1.81
1900	I Baxter	USA	1.90
1904	S Jones	USA	1.80
1906	C Leahy	GBR	1.78
1908	H Porter	USA	1.91
1912	A Richards	USA	1.93
1920	R Landon	USA	1.94
1924	H Osborn	USA	1.98
1928	R King	USA	1.94
1932	D McNaughton	CAN	1.97
1936	C Johnson	USA	2.03
1948	J Winter	AUS	1.98
1952	W Davis	USA	2.04
1956	C Dumas	USA	2.12
1960	R Schavlakadze	URS	2.16
1964	V Brumel	URS	2.18
1968	D Fosbury	USA	2.24
1972	J Tarmak	URS	2.23
1976	J Wszola	POL	2.25
1980	G Wessig	GDR	2.36
1984	D Mogenbury	GER	2.35

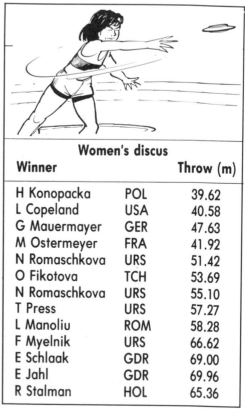

Women's discus

Winner		Throw (m)
H Konopacka	POL	39.62
L Copeland	USA	40.58
G Mauermayer	GER	47.63
M Ostermeyer	FRA	41.92
N Romaschkova	URS	51.42
O Fikotova	TCH	53.69
N Romaschkova	URS	55.10
T Press	URS	57.27
L Manoliu	ROM	58.28
F Myelnik	URS	66.62
E Schlaak	GDR	69.00
E Jahl	GDR	69.96
R Stalman	HOL	65.36

Olympic medals

Information ⮡ At each Olympic Games a medals table is drawn up showing the total number of gold, silver and bronze medals each country wins. The countries are ranked according to the number of gold medals they have won, followed by silver and bronze. There are other ways that the countries could be ranked.

Investigations

1 Below is the medals table for the 1976 Olympic Games. Award 3 points for each gold medal, 2 points for each silver medal and 1 point for each bronze medal. Count up the total number of points for each country. How would this affect the order of the 1976 table? Reorder the table using this method to find the top 10 countries.

Country	Gold	Silver	Bronze	Points	Population (millions)
USSR	47	43	35	262	256.7
E Germany	40	25	25		16.8
USA	34	35	25		215.1
W Germany	10	12	17		61.5
Japan	9	6	10		112.8
Poland	8	6	11		34.4
Bulgaria	7	8	9		8.8
Cuba	6	4	3		9.5
Romania	4	9	14		21.4
Hungary	4	5	12		10.6
Finland	4	2	0		4.7
Sweden	4	1	0		8.2
UK	3	5	5		55.9
Italy	2	7	4		56.2
Yugoslavia	2	3	3		21.6
France	2	2	5		52.9
Czechoslovakia	2	2	4		14.9
New Zealand	2	1	1		3.1
Switzerland	1	1	2		6.3
S Korea	1	1	2		35.9

2 Countries like America and USSR have large populations and therefore many athletes to choose from for their teams. Smaller countries do not have the same numbers of athletes. One way to balance this out is to divide the points total for each country by the population in millions to find the points per million of population. How would the medals table for 1976 be altered using this method? Reorder the table using this method to show the top ten countries.

Country	Gold	Silver	Bronze	Points	Population (millions)	Points per million
USSR	47	43	35	262	256.7	1.02
.	.	.	.	.	.	.
.	.	.	.	.	.	.
.	.	.	.	.	.	.

Why is this method fairer? Can you think of any other fair way of producing a medals table? Try it out and see if it changes the order.

Averages

Information ⇨ It is quite common in sports, games and competitions to work out averages in order to place teams in leagues or players in world ranking. The correct name for the average calculated in this way is the **mean**.

Example

Calculating the mean from a list of data
The list below shows the runs scored by each batsman in a local cricket match. What is the mean number of runs scored by each batsman?

5, 6, 6, 3, 8, 5, 4, 5, 1, 7, 2, 3, 7, 5, 8, 4, 1, 7, 4, 2, 3, 1

To find the average or mean you add up all the runs and divide by 22 (the total number of batsmen).

$(5+6+6+3+8+5+4+5+1+7+2+3+7+5+8+4+1+7+4+2+3+1) \div 22$

$97 \div 22 = 4.41$ runs

THAT'S A MEAN SCORE YOU'VE GOT!

Example

Calculating the mean from a frequency table
A spectator at the same cricket match listed the runs in a frequency table. Find the mean number of runs using the information in the frequency table.

To calculate the mean directly from the table, you must calculate $f \times x$ for each row in the table, as shown. Adding up this new column, fx, will give you the total runs scored — 97 as above.

Mean is $97 \div 22 = 4.41$

To calculate the mean from a frequency table you can use the formula

Mean = fx total ÷ f total

Runs (x)	Frequency (f)	fx
0	0	0
1	3	3
2	2	4
3	3	9
4	3	12
5	4	20
6	2	12
7	3	21
8	2	16
Totals	22	97

You get the same mean number, 4.41, whichever method is used to calculate the mean.

Exercise 4   Copy and complete each of the frequency tables in questions **1** to **4**. Calculate the mean in each case.

1 The number of goals scored by 12 football teams one Saturday.

Goals (*x*)	Frequency (*f*)	*fx*
0	2	0
1	3	3
2	1	
3	2	
4	3	
5	1	
Totals		

How many goals were scored in total?

2 The ages of girls who play volleyball in an under 16 competition.

Age (*x*)	Frequency (*f*)	*fx*
11	5	55
12	6	72
13	9	
14	12	
15	16	
Totals		

There are 6 players in each team. How many teams were in the competition?

3 The scores in a golf competition.

Score (*x*)	Frequency (*f*)	*fx*
78	1	78
79	3	
80	5	
81	7	
82	5	
83	0	
84	4	
85	2	
86	1	
Totals		

How many people took part in the competition?

4 The ages of the players in a junior hockey league.

Age (*x*)	Frequency (*f*)	*fx*
13	16	208
14	17	
15	29	
16	38	
17	43	
Totals		

There are 11 players in a hockey team. How many teams entered the league?

5 The average attendance at a football match ground was 25 000. Due to a good run of results the average attendance increased by 8%. What was the new average attendance?

6 Carol is training to run the London marathon. She can keep up a steady pace and runs each mile in an average time of 9½ minutes. How long will it take her to run the marathon which is 26.22 miles long?

7 Raschid is also training for the London marathon. His target time is 3 hours. What must be his average time for each mile to achieve this?

8 The average attendance at a cricket ground was 820. Due to a spell of bad weather, the average attendance dropped by 7%. What was the new average attendance?

9 When hill walking, Peter averages a distance of 4 km every hour. How long will it take him to walk 15 km?

Information ⇨ If there is a big range in the data being measured it is easier to **group** the data in a frequency table. The groups are arranged so that no score is repeated in adjacent groups. The **mid-point** of each group is used to calculate the mean.

Examples

Mean from grouped frequency

1 Here are the first round scores in a shooting competition. Find the mean score. For each range of scores the mid-point (x) and the value of fx have been calculated for you here.

Score	Mid-point (x)	Frequency (f)	fx
1−5	3	1	3
6−10	8	3	24
11−15	13	5	65
16−20	18	6	108
21−25	3	4	92
26−30	28	3	84
31−35	33	1	33
Totals		23	409

Mean = fx total ÷ f total
= 409   ÷   23   =   17.8

2 A survey recorded the speeds of skiers on a downhill run. The results are shown in the table below. Find the mean speed.

Speed (kph)	Mid-point (x)	Frequency (f)	fx
21−30	25.5	5	127.5
31−40	35.5	10	355.0
41−50	45.5	3	91.0
51−61	55.5	2	111.0
61−70	65.5	1	65.0
Totals		20	749.5

Mean = fx total ÷ f total
= 749.5   ÷   20   =   37.5 kph.

Exercise 5 Find the mean for each of the frequency tables of grouped data in questions **1** to **4**.

1 The scores in a day's golf on a local course.

Score	Mid-point (x)	Frequency (f)	fx
66−70	68	0	0
71−75	73	1	73
76−80	78	3	
81−85	83	19	
86−90		18	
91−95		26	
96−100		10	
Totals			

2 The length of time taken for snooker games.
Find the mean correct to the nearest 1 second.

Time (min)	Mid-point (x)	Frequency (f)	fx
16—20	18	3	54
21—25	23	4	
26—30		6	
31—35		4	
36—40		2	
41—45		1	
Totals			

3 The attendance at a local amateur football teams home games.

Attendance	Mid-point (x)	Frequency (f)	fx
1—20	10.5	3	
21—40		5	
41—60		7	
61—80		3	
81—100		2	
Total			

4 The scores per throw over a set of darts games.

Score	Mid-point (x)	Frequency (f)	fx
1—30	15.5	3	46.5
31—60	45.5	5	227.5
61—90	75.5	8	
91—120		4	
121—150		2	
151—180		2	
Totals			

5 The runs scored by each player in a cricket match.

Runs	Mid-point (x)	Frequency (f)	fx
1—25	13	4	
26—50		7	
51—75		5	
76—100		4	
101—125		2	
Totals			

Probability

Information ⇨ You have probably heard people saying things like; "go on, take a chance", or "I wouldn't bet on it". These expressions are about the chance of something happening.

Probability is a measure of chance. An event that is certain to happen has a probability of 1. An event that is certain not to happen has a probability of 0. All other probabilities lie between 0 and 1.

To calculate the probability of something happening you need to know the number of **favourable** results and the number of **possible** results. A favourable result is one which you want to happen.

$$\text{Probability} = \frac{\text{number of favourable results}}{\text{number of possible results}}$$

Examples **Calculating a probability**

1 What is the probability of getting a head if you toss a coin?

Number of favourable results = 1 (there is 1 head on a coin)
Number of possible results = 2 (head or tail)
Probability of getting a head is ½

2 What is the probability of picking a heart out of a pack of cards?

Number of favourable results = 13
Number of possible results = 52
Probability of picking a heart is $\frac{13}{52} = \frac{1}{4}$

Exercise 6 In questions **1** to **5** write down the probability of:

1 getting a tail if you toss a coin.

2 picking a red card from a pack of cards.

3 getting a six if you roll a die.

4 picking a face card from a pack of cards.

5 getting an even number if you roll a die.

6 Vijay bought 4 raffle tickets. 200 tickets were sold altogether. What is Vijay's chance of winning?

7 Here is a probability scale. Copy it and put the events in questions **1-6** above in the correct order on the scale. Think of some more of your own.

No chance   Probability of getting a head   Certainty

0   ½   1

Information ☞ You can try some experiments to see if your estimates for probability are correct. In each of the experiments below you should try to estimate the outcome before doing the experiment.

Example

Coin experiment
Jill and Dave tossed a coin 50 times and recorded their results in a table. The probability of getting a head is ½ so they expected to get 25 heads. Here are their results: Heads 23.   Tails 27.

The experimental results are quite close to the expected outcome.

Exercise 7

In this exercise you should first try to estimate the results of the experiments. After doing the experiment you should check to see if your estimate was close to the experimental result.

1   Try the same coin experiment as Jill and Dave but this time do it for 100 tosses. How many Heads would you expect to get? How many did you get?

2   Roll a die 60 times. How many sixes would you expect to get? How many did you actually get?

3   Get a pack of cards. Pick a card and record its suit (the type of card) in a table like the one shown. Repeat this 100 times.

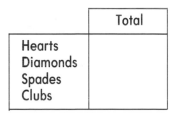

	Total
Hearts	
Diamonds	
Spades	
Clubs	

(a) How many clubs would you expect to pick?
(b) How many red cards would you expect to pick? (Remember Diamonds and Hearts are all red cards.)

4   Get a shallow box and 50 drawing pins. Throw the pins into the box. How many pins would you expect to land point up. Do the experiment. Does your result surprise you?

Joint events

Information ⇨ The probability of events happening can be combined to give the probability of both events happening. One way of finding the probability of two events happening is to multiply the individual probabilities together.

Example

Probability of two events
What is the probability of getting two heads if two coins are thrown?

Possible outcomes are

Coin 1	Coin 2	
T	T	
T	H	
H	T	
H	H	(favourable)

Number of favourable outcomes = 1
Number of possible outcomes = 4
Probability = ¼

Check by individual probabilities
Probability of a head × Probability of a head = ½ × ½ or ¼

Exercise 8

1   Copy and complete this table showing the possibilities when two dice are thrown.

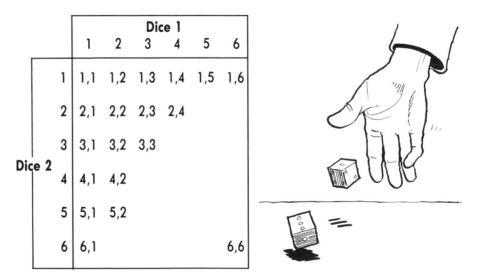

		Dice 1					
		1	2	3	4	5	6
Dice 2	1	1,1	1,2	1,3	1,4	1,5	1,6
	2	2,1	2,2	2,3	2,4		
	3	3,1	3,2	3,3			
	4	4,1	4,2				
	5	5,1	5,2				
	6	6,1					6,6

2   How many possible outcomes are there in the table? How many outcomes are there for a double six? What is the probability of getting a double six?

3   What is the probability of getting a double three?

4   What is the probability of getting a total of seven?

5   What is the probability of getting a total more than 9?

6   What is the probability of getting a difference of 3?

7   What is the probability of getting a head followed by a tail when tossing a coin twice?

8   What is the probability of picking a Heart then a Club from a pack of cards (card is returned to the pack after each choice)?

9   You have a coin and a dice. What is the probability of
(a) a Head followed by a six?
(b) a Head followed by a four?
(c) a Tail followed by an even number?
(d) a Tail followed by an odd number?

10   You have a pack of cards and a coin. What is the probability of
(a) an Ace followed by a Head?
(b) a face card followed by a Tail?
(c) a Diamond followed by a Head?
(d) a red card followed by a Tail?

Investigations

Dice Bingo
You need a partner and two dice to play 'Dice Bingo'.

You write down 4 numbers between 2 and 12. You can repeat a number if you wish.
Your partner rolls the dice, adds the scores and calls out the total.
If you have that number you cross it off.
Only 1 number can be crossed off per throw.
How many calls are made before you have crossed out all your numbers?

How could you increase your chances?

What difference would it make if you (a) subtracted the dice scores?
(b) multiplied the dice scores?

Six sixes
In some fetes and fund-raising events there are competitions to win big prizes like a car. All you have to do is throw six sixes with six dice! What do you think your chances will be?

Permutations

Information ↩ The ways in which a set of letters or a list of features can be arranged are called **permutations**. The number of permutations for arranging a group depends on the number of items in the group.

Example

Investigating permutations
Alan collected a number of different coloured counters.

With 1 item there is only one way to arrange it. There is 1 permutation.

How many difference ways can Alan arrange 2 counters?

Alan took a red and a blue counter. He arranged them as (red,blue) or (blue,red). So there are 2 ways to arrange 2 different coloured counters.

There are 2 permutations.

Investigation

You will need 4 different coloured counters to help you with this investigation.

1 Take 3 counters, for example 1 red, 1 blue and 1 green. List all the different ways you can arrange the counters.
Here are the first 4 arrangements:
(red, blue, green) (red, green, blue) (blue, red, green) (blue, green, red)
Copy these 4 answers and find the others. How many permutations are there for 3 counters?

2 Use an additional coloured counter to the 3 you used in question **1**, for example, black. How many different ways can you arrange 4 counters? Here are
4 arrangements:
(red, blue, green, black) (red, blue, black, green)
(red, green, blue, black) (red, green, black, blue).
Copy these 4 and list all the other possible arrangements. How many permutations are there for 4 counters?

3 Copy and complete the table.

Number of counters   Permutations

```
1 — — — — 1
2 ⇄ → 2
3 ⇄ → 6
4 ⇄ → 24
5 —
```

The dotted lines show how the table is built up.
$1 \times 2 = 2$ $2 \times 3 = 6$ $6 \times 4 = 24$ and so on . . .

New car competition

Here is Dermot's entry for the competition.
Does his order agree with yours?
Does it agree with the order chosen by
anyone in your class?

WIN A NEW CAR

Listed below are 10 important features of the new FIAT PANDA range

a	Good fuel economy	6
b	Dual circuit brakes	1
c	Fold-down rear seat	9
d	Five-speed gearbox	10
e	Single overhead camshaft	2
f	Good all round vision	7
g	Rear screen wiper and washer	4
h	Tinted glass	3
i	Sun roof	8
j	Front-wheel drive	5

Put these features in order of importance

Dermot had to arrange 10 times. To work
out his chances of winning you need to
know how many different permutations
there are for the 10 items to be
arranged in order.

Part of the solution is given in the table
in investigation **3** on the previous page.
Here is the table. Do you remember the
link between one number and the next?

Number of items	Permutations
1	1
2	2
3	6
4	24
5	120
6	720
8	
9	
10	

Copy and complete the table. A
calculator will help.

Now do you think Dermot has much chance
of winning the car?

Investigations

1 Here are 2 investigations which give unexpected results.

(a) There are 365 days in a year and people can have birthdays on any day. Ask 30 people to tell you the date of their birthday this year. Note the results.

(b) There are 100 two digit numbers from 00 to 99. Stand on the pavement at the roadside and write down the last 2 digits on the number plates of 20 vehicles as they pass. For example if the number is B 631 AHS you would write down 31. Are there 2 the same? Compare your results with others in your class.

2 A 400 metre running track is made up of 8 lanes each 1.8 metres wide. The distance round the inside lane is 400 metres.

How far will a runner in lane 2 cover in 1 circuit of the track? How is this problem overcome? What happens on the other lanes?

4 In darts each player must finish with
a double, or a bull which scores 50.
For example if a player is left with
16 he would try to get a double 8.
He has 3 darts to do this and can
score a single, double or triple of
any number between 1 and 20.

What scores are needed to check out these numbers with 3 darts or less?
(**a**) 101 (**b**) 140 (**c**) 139 (**d**) 134
(**e**) What is the highest number that can be checked out with 3 darts?
(**f**) Make a list of the numbers between 150 and 170 which cannot be checked out
with 3 darts.

5 Design, and draw out in the playground, a running track 100 m long. It should be
made up of 4 equal sections, 2 parallel straights and 2 semi-circles at either end.
Once you have drawn the track check your accuracy with a trundle wheel.
Use your 100 m track to find out your average length of pace. How does this
compare with the class average?

Is there much difference between the average pace of the girls and boys?

6 Bob, Carol, Ted, Alice, Senga and Ryland all go to the cinema and sit together in 1
row. How many different seating arrangements are possible?

eg Bob, Carol, Ted, Alice, Senga, Ryland;
or Bob, Carol, Ted, Alice, Ryland, Senga

Hint: Solve the seating problem for 2 people, then 3 people, then 4 people. Try to
draw conclusions from your results and use them to solve the problem.

7 The next week, all 6 go to a concert together but agree that everyone must sit
beside at least 1 person of the opposite sex. How many different seating
arrangements are possible this time.

11 Right-angled triangles

Do you remember? ⟿ In this unit it is assumed that you know how to:

1 express numbers to a given degree of accuracy

2 solve simple equations involving multiplication or division.

Exercise 0 will help you check. Can you answer all the questions?

Exercise 0 **Accuracy**

1 Write these numbers to the nearest 10.
(a) 52 (b) 81 (c) 28 (d) 47 (e) 129 (f) 263

2 Write these numbers to the nearest 100.
(a) 87 (b) 135 (c) 678 (d) 215 (e) 525 (f) 1478

3 Write these numbers correct to 1 decimal place.
(a) 2.71 (b) 6.39 (c) 54.781 (d) 148.649

4 Write these numbers correct to 1 significant figure.
(a) 63 (b) 51 (c) 89 (d) 127 (e) 37.5 (f) 49.87

5 Write these numbers correct to 2 significant figures.
(a) 237 (b) 666 (c) 1987 (d) 41.23 (e) 7.654

Solving simple equations
Solve these equations to find the value of x.

6 $\frac{x}{2} = 4$ 7 $\frac{x}{2} = 6$ 8 $\frac{x}{2} = 12$

9 $\frac{x}{3} = 2$ 10 $\frac{x}{3} = 5$ 11 $\frac{x}{3} = 7$

12 $\frac{x}{5} = 2$ 13 $\frac{x}{7} = 8$ 14 $\frac{x}{4} = 12$

The next ones are slightly different.

15 $\frac{6}{x} = 2$ 16 $\frac{8}{x} = 2$ 17 $\frac{10}{x} = 2$

18 $\frac{15}{x} = 3$ 19 $\frac{12}{x} = 3$ 20 $\frac{21}{x} = 3$

21 $\frac{16}{x} = 8$ 22 $\frac{20}{x} = 4$ 23 $\frac{45}{x} = 5$

A problem of access

Mike and Kirsten both help out at their local Community Centre. The PHAB (Physically Handicapped Able Bodied) club want to use the facilities but have problems getting wheelchairs into the centre because of the steps outside. Mike and Kirsten have been asked to look into the possibility of building a ramp like the one shown below.

The steps climb a total height of 80 cm.
The ramp must not be steeper than 20° from the horizontal.

How long will the ramp be ?
How far back from the steps will the ramp come?

This unit will explain how to use trigonometry to solve problems like this.

Sides and angles

Information ➭ In this unit we are going to look at **trigonometry.** In particular we are going to examine the relationship between the length of the sides and the size of the angles in right-angled triangles.

Investigation   1   Draw out the following right-angled triangles accurately.

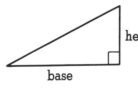

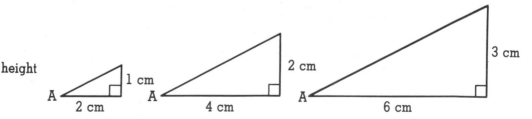

2   Using information from the triangles you have drawn copy and complete the table below. Take the measurements as accurately as possible. Give answers in the last column correct to 1 decimal place.

Angle at A	Base	Height	Height ÷ Base
26°	2	1	0.5

Write down any conclusions you can draw from the table.

Does the same thing happen in all similar right-angled triangles?

3   Draw out the following right-angled triangles accurately. Continue the pattern until you have drawn another 4 similar right-angled triangles. Copy and complete the table below using the information from the triangles you have drawn.

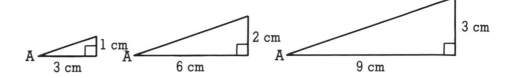

Angle at A	Base	Height	Height ÷ Base

4   Repeat question **3** for a different series of similar right-angled triangles.

5   Do all your results show the same thing?

Labelling sides

Information ✏   When you use trigonometry to solve problems it is important to label diagrams clearly and correctly.

Example   **Labelling sides of a right-angled triangle**

Here is a right-angled triangle with angle BÂC marked $x°$.

AB is opposite the right-angle and is called the **hypotenuse.**

BC is facing angle x and is called the **opposite** side.

AC is beside angle x and is called the **adjacent** side. The sides are labelled **h**, **o** and **a.**

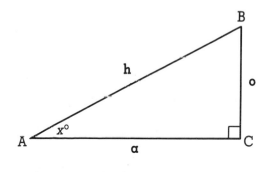

Exercise 1   Copy each of these right-angled triangles. The right-angle and angle $x°$ have been labelled. Label the sides **h** for hypotenuse, **o** for opposite and **a** for adjacent.

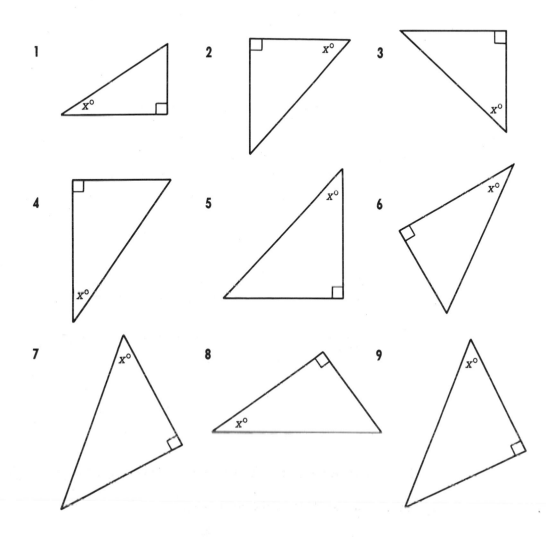

Tangent

Information ⤸ The results from the investigation on page 195 show you two things.

(a) In each series of similar triangles the angle at A is the same.
(b) In each series of similar triangles height ÷ base is the same.

The ratio   height : base   is therefore a measure of angle A.

This ratio is called the **tangent** of angle A (**tan A**).

The tangent ratio can be used to calculate the size of an angle. The tangent of angles between 0° and 90° are found either by looking up trigonometry (trig) tables, or by using a scientific calculator.

Example   **Finding tan values**
Find tan 26.7°.

Using tables
Look for Natural Tangents in the tables
Look down column until you reach 26°
Go along to .7 and read tan value
tan 26.7° = 0.503

Using calculator
Enter angle size 26.7°
Press tan button
Read tan value = 0.503
(Round up to 3 decimal places)

Natural Tangents

Degrees	.0	.1	.2	.3	.4	.5	.6	.7	.8	.9
25	0.466	468	471	473	475	477	479	481	483	486
26	.488	499	492	494	496	499	501	**503**	505	507
27	.510	512	514	516	518	521	523	525	527	529

Example   **Finding the angle given the tangent**
Find the angle which has a tan of 1.582.

Look for 1.582 in main part of table. Read values from left column and top line.
Angle is 57.7°

You can also do this on a scientific calculator. Ask your teacher to show you how. (Note that the tangent of an angle greater than 45° is greater than 1.)

Natural Tangents

Degrees	.0	.1	.2	.3	.4	.5	.6	.7	.8	.9
56	1.428	433	439	444	450	455	460	466	471	477
56	.483	488	494	499	505	511	517	522	528	534
57	.540	546	552	558	564	570	576	**582**	588	594

Exercise 2   Use the tables, or a scientific calculator to find

1 tan 32.5°   **2** tan 45°   **3** tan 67.8°   **4** tan 86.2°

5 Angle which has tan of 0.475   **6** Angle which has tan of 1.921

Information ⇨ You can use the tangent ratio to calculate the size of an unknown angle in a right-angled triangle.

Examples **Calculating the size of an angle**

Calculate the size of angle $x°$.

Mark the hypotenuse, opposite and adjacent sides.

$$\tan = \frac{\text{opposite}}{\text{adjacent}}$$

$$\tan x° = \frac{9}{13}$$

$$\tan x° = 0.692 \text{ (to 3 significant figures)}$$

From the tables the angle whose tangent is 0.692 is 34.7°.

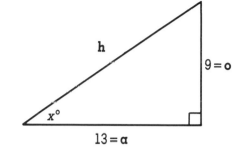

Exercise 3 **Use the tangent ratio to find angle** $x°$ **in each of these right-angled triangles.**

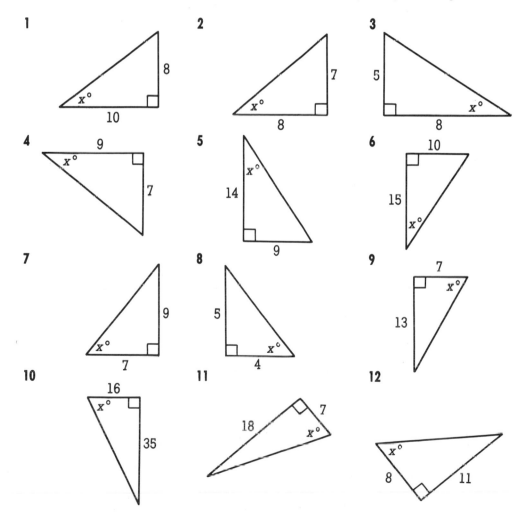

Sine, cosine and tangent

Information ↩   You have seen how the opposite and adjacent sides of a right-angled triangle give the tangent ratio and how this can be used to find the size of angles in right-angled triangles. There are two other ratios you can use. The **sine** ratio is given by the opposite and hypotenuse. The **cosine** ratio is given by the adjacent and hypotenuse.

In this right-angled triangle one angle is marked *x*. The hypotenuse, opposite and adjacent sides are also marked.

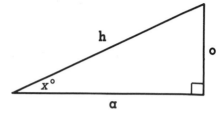

SOH CAH TOA helps to remind us of the 3 ratios and sides.

$$\sin = \frac{\text{opposite}}{\text{hypotenuse}} \qquad \cos = \frac{\text{adjacent}}{\text{hypotenuse}} \qquad \tan = \frac{\text{opposite}}{\text{adjacent}}$$

Examples   **Using sine, cosine and tangent ratios to calculate angles**
Find angle *x*° in each of these right-angled triangles.

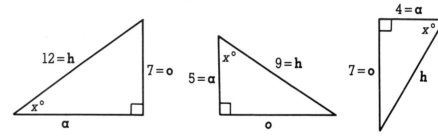

Mark the sides **h**, **o** and **a** in each case.

SOH	CAH	TOA
o and **h** are known so sin	**a** and **h** are known so cos	**o** and **a** are known so tan
$\sin = \frac{o}{h}$	$\cos = \frac{a}{h}$	$\tan = \frac{o}{a}$
$\sin x° = \frac{7}{12}$	$\cos x° = \frac{5}{9}$	$\tan x° = \frac{7}{4}$
$\sin x° = 0.583$	$\cos x° = 0.556$	$\tan x° = 1.750$
$x = 35.7°$	$x = 56.2°$	$x = 60.3°$

Exercise 4   Use the sine ratio to find angle *x*° in these right-angled triangles.

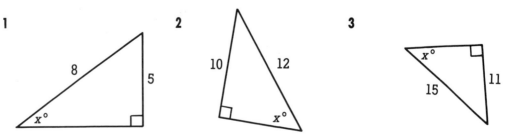

Use the cosine ratio to find angle x° in these right-angled triangles.

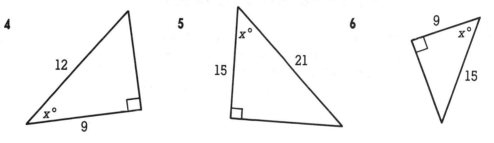

4

5

6

Use the tangent ratio to find angle x° in these right-angled triangles.

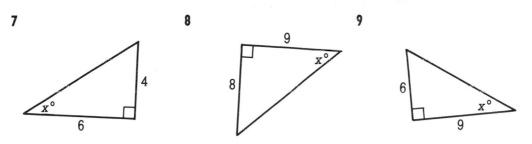

7

8

9

Calculate angle x° in these right-angled triangles. Decide which ratio you must use.

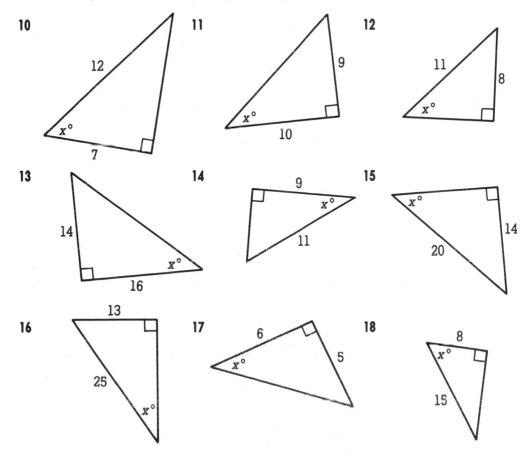

10

11

12

13

14

15

16

17

18

Finding the length of sides

Information ⇨ You have seen how to find the size of angles in right-angled triangles. Now you are going to look at the problem of finding the length of sides.

Examples

Using sine, cosine and tangent ratios to find the length of sides
Find sides p, q and r.

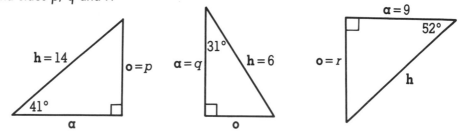

Mark the sides **h**, **o** and **a** in each case.

SOH
o and h are known
so use sin

$\sin 41° = \dfrac{p}{14}$

$0.656 = \dfrac{p}{14}$

$p = 14 \times 0.656$
$p = 9.18$

CAH
a and h are known
so use cos

$\cos 31° = \dfrac{q}{6}$

$0.857 = \dfrac{q}{6}$

$q = 6 \times 0.857$
$q = 5.14$

TOA
o and a are known
so use tan

$\tan 52° = \dfrac{r}{9}$

$1.280 = \dfrac{r}{9}$

$r = 9 \times 1.280$
$r = 11.5$

Give answers to 3 significant figures.

Exercise 5

Use the sine ratio to find side p in these right-angled triangles.

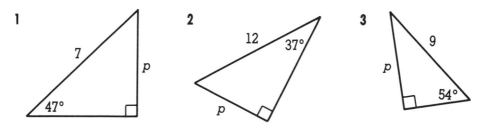

Use the cosine ratio to find side q in these right-angled triangles.

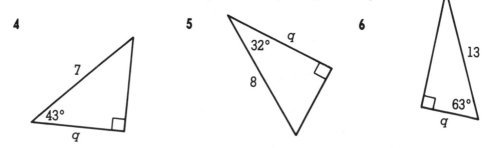

Use the tangent ratio to find side *r* in these right-angled triangles.

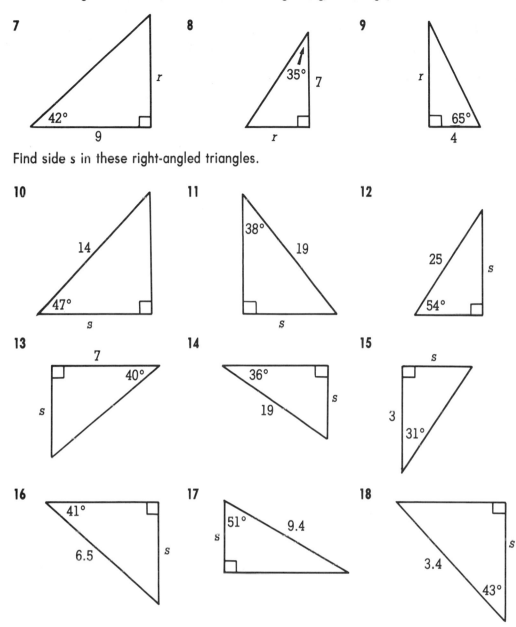

7

42°

9

r

8

35°

7

r

9

r

65°

4

Find side *s* in these right-angled triangles.

10

14

47°

s

11

38°

19

s

12

25

s

54°

13

7

40°

s

14

36°

19

s

15

s

3

31°

16

41°

6.5

s

17

51°

9.4

s

18

3.4

43°

s

A sketch will help you solve the following problems.

19 A ladder is 5 metres long and rests against the side of a house. The foot of the ladder makes an angle of 55° with the level ground.

(**a**) How high up the wall does the ladder reach?
(**b**) How far is the foot of the ladder from the wall?

20 A telegraph pole, standing on level ground, is 9 metres tall and is supported by 2 wire guys attached to the top of the pole. Each guy makes an angle of 28° with the top of the pole. How far away from the foot of the pole are the guys anchored?

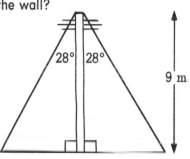

28° 28°

9 m

Information ⟿ So far you have multiplied to find the length of a side in a right-angled triangle.
Sometimes you need to use division to work out the length of a side using the ratios.

Examples   **Calculating the hypotenuse and adjacent sides using division**
Find sides *p*, *q* and *r* in these right-angled triangles.

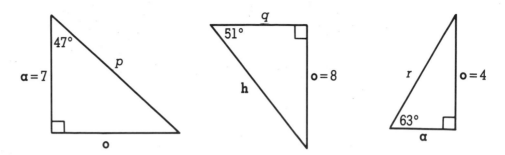

Mark the sides **h**, **o** and **a** in each case.

p=**h**	*q*=**a**	*r*=**h**
CAH	TOA	SOH
a is known and	**o** is known and	**o** is known and
h is wanted so cos	**a** is wanted so tan	**h** is wanted so sin

$$\cos 47° = \frac{7}{p}$$

$$0.682 = \frac{7}{p}$$

$$0.682p = 7$$

$$p = \frac{7}{0.682}$$

$$p = 10.3$$

$$\tan 51° = \frac{8}{q}$$

$$1.235 = \frac{8}{q}$$

$$1.235q = 8$$

$$q = \frac{8}{1.235}$$

$$q = 6.48$$

$$\sin 63° = \frac{4}{r}$$

$$0.891 = \frac{4}{r}$$

$$0.891r = 4$$

$$r = \frac{4}{0.891}$$

$$r = 4.49$$

Exercise 6   Find the length of side *s* in each of these right-angled triangles.

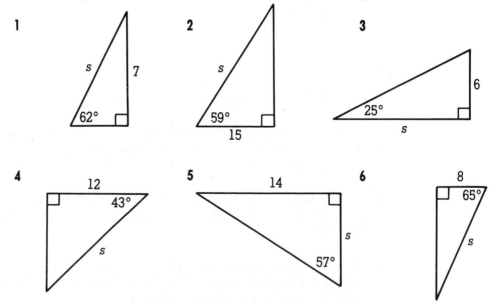

The following triangles are a mixture. In some you have to divide to find the missing sides while in others you have to multiply.

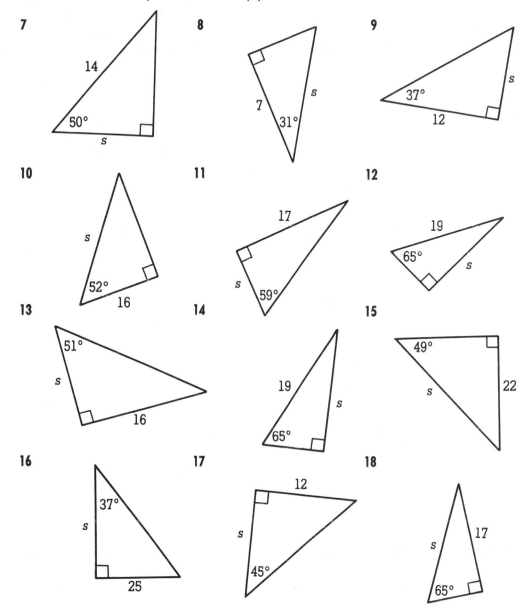

A sketch will help you solve the following problems.

19 A flag pole, 4 metres high, is held in place by 3 wire guys. Each guy is attached to the top of the pole and anchored to the level ground making an angle of 68°. How long is each guy?

20 A boy is flying a kite at a height of 28 metres. The string is taught and makes an angle of 68° to the horizontal. How much string has he wound out to get the kite up to this height?

21 An aircraft takes off and climbs at an angle of 35° to the horizontal until it reaches a height of 5000 metres. How far will the aircraft have travelled towards its destination by the time it reaches this height?

A problem of access

Below is a sketch showing the steps outside Mike and Kirsten's local Community Centre.

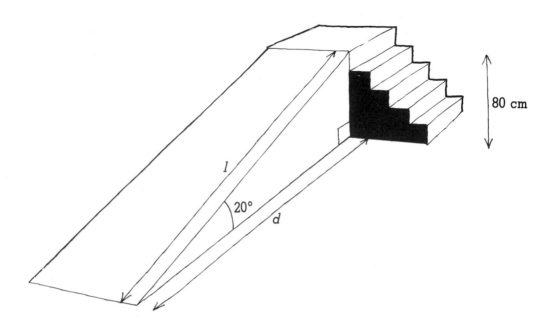

From the sketch a right-angled triangle is drawn.

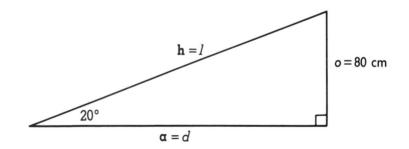

Length of the ramp *l* (**h**):

SOH
o is known and **h** is wanted so use sin

$$\sin 20° = \frac{80}{l}$$

$$0.342 = \frac{80}{l}$$

$$0.342l = 40$$

$$l = \frac{80}{0.342}$$

$$l = 234 \text{ cm}$$

The distance from the steps *d* (**a**):

TOA
o is known and **a** is wanted so use tan

$$\tan 20° = \frac{80}{d}$$

$$0.364 = \frac{80}{d}$$

$$0.364d = 80$$

$$d = \frac{80}{0.364}$$

$$d = 220 \text{ cm}$$

The ramp is 234 cm long and will be 220 cm from the steps.

Investigations

1   On a building site rubbish is put into a skip. A plank is used to give access to the skip for wheelbarrows. The plank must not be steeper than 15° to the horizontal. The edge of the skip is 1.2 metres high.

What is the minimum length of plank needed? How far out from the edge of the skip will the plank touch the ground?

2   The sketch below shows a child's chute. It is designed for 3 year old children. The slide should be at an angle of 55° to the horizontal and the steps at an angle of 65° to the horizontal.

What is the total length of the chute including steps and slide?

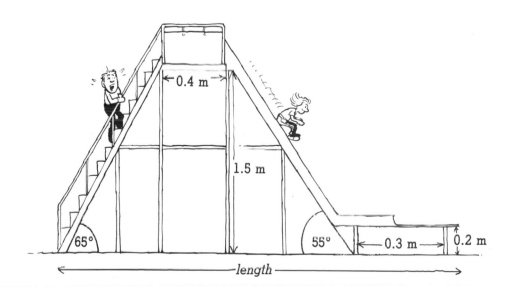

12 Money and the home

Do you remember? ↩ In this unit it is assumed that you know how to:

1   multiply using common fractions
2   find a percentage of a sum of money
3   work out problems using a rate
4   change between weekly, monthly and annual costs.

Exercise 0 will help you check. Can you answer all the questions?

Exercise 0   **Multiplication using common fractions**

1 $4000 \times 2\frac{1}{2}$	**2** $6400 \times 1\frac{1}{4}$	**3** $2000 \times 2\frac{1}{4}$
4 $3800 \times 1\frac{3}{4}$	**5** $7200 \times 3\frac{1}{4}$	**6** $5200 \times 2\frac{3}{4}$

Find a percentage of a sum of money

7 Find 90% of £7000.   **8** Find 85% of £19 000.

9 Find 95% of £45 500.   **10** Find 90% of £52 250.

11 Find 85% of £38 750.   **12** Find 95% of £42 200.

Problems using a rate

13 A generator uses fuel at the rate of 1 gallon every 3 hours. How much fuel is needed to keep the generator running constantly for 1 week?

14 Jackie is saving up to buy a motor bike. Her mother agrees to add £20 to every £100 she saves. How much does her mother owe her if Jackie saves up £625?

Weekly, monthly and annual costs

15 Sajid gets £3.50 per week pocket money. How much is this per year?

16 Senga pays £39 per month to a mail order company. How much is this per year? She changes to weekly payments. How much will she pay each week?

17 Moira saves £3.50 per week. Sheena saves £14 per month. Who will save more in a year?

18 Bill earns £545 per month and Brian earns £128 per week. Who earns more in a year?

19 Fiaz gets £4 per week pocket money. Ihab gets £18 per month pocket money. Who is better off?

Flat-mates

Sue and Angela are moving away from home. They both work in a new engineering factory which is 10 miles from the town in which they live. Susan is an engineer and will earn £8600 per year. Angela is a secretary and will earn £545 per month. They are thinking about sharing a flat together.

Should they rent or buy?

Can they afford all the bills?

Could they get a mortgage to pay for a flat?

How much would a mortgage cost?

Sue and Angela have got a lot to think about. This unit will help you advise them about all the things they should consider.

Renting a place

Information ⇨ If you are thinking about leaving home for the first time it is a good idea to work out if you can afford to live away from home. There are many costs to consider. For example rent, rates, food, electricity, gas, clothes etc. Some of these have to be budgeted for, others are luxuries. **Rent** is a major cost. Sometimes **rates** are included in the rent but sometimes they are charged separately.

Example   **Calculating weekly rent**
Three friends decide to share a flat. They see this advert. Work out the rent and rates each person will have to pay assuming that they pay equal amounts.

SPACIOUS FLAT in large Victorian house. Ideal for 3 to share. 1 large twin bedroom, 1 single bedroom, living room, kitchen bathroom. £311 per month plus £40 rates

Total rent and rates:	£311+£40 = £351
Total for year:	£351×12 = £4212
Weekly total:	£4212÷52 = £81
Each person's share:	£81÷3 = £27

Exercise 1   Look at these adverts. Choose the ones you think might be suitable for 2 people sharing a flat.

For each advert you choose write down how much it would cost each person per week to rent the flat.

1 **CENTRAL** flat, 2 bedrooms, lounge, kitchen, bathroom. £280 per month.

2 **ROOMS** in large flat. 2 rooms in central flat £162 per month.

3 **QUIET** flat for rent. 1 double 1 single bedroom, lounge/kitchen, WC, shower room. Immediate entry. £265 per month.

4 **COMPACT** flat comprising living room/kitchen, 2 bedrooms bathroom £254 per month plus £36 rates.

5 **SINGLE** room in flat. Professional person, non-smoker. £150 per month plus £23 rates.

6 **CENTRAL** flat comprising lounge, kitchenette, 2 bedrooms toilet, shower room £310 per month.

Household bills

Information ✍ You can try to estimate what some of the main bills may be in a new flat or house by looking at a set of household bills for the past few months. These bills are usually paid every 3 months.

Example

Estimating household bills

The Mackenzie family, Janet, Hugh, Ian and Dave, paid these amounts for bills between January and June.

January	Gas £75.42 Electricity £89.94 Food £180 TV £17
February	Phone £32.93 Food £194 TV £17
March	Food £178 TV £17
April	Gas £86.92 Electricity £49.98 Food £190 TV £17
May	Phone £42.70 Food £165 TV £17
June	Food £175 TV £17

What is the average cost of the family's electricity consumption per week?

Total electricity bills   £89.94 + £49.98 = £139.92
January to June is half a year or 26 weeks
Average weekly electricity cost £139.92 ÷ 26 = £5.38

Exercise 2

1   Calculate the average weekly cost for the Mackenzie family for (**a**) gas, (**b**) food, (**c**) telephone and (**d**) television.

2   Ian and Dave Mackenzie have just finished college and decide to rent a small flat. They decide to do without a telephone or television to save money. They also think that, because they will be in a smaller place than their parents' house, they can knock 20% off each of the remaining costs.
How much are they each going to have to budget per week for (**a**) gas, (**b**) electricity and (**c**) food?

3   The flat they have rented costs £245 per month. How much per week must they each put aside for rent?

4   Between them they earn £194 per week. Use the answers to questions **2** and **3** to work out how much they will have left to spend on entertainment, clothes, records, etc. between them.

Buying a flat

Information ➪ If you are thinking about buying a house or flat you will probably have to borrow a lot of money, unless you are extremely rich! Most people borrow money from a building society or bank. This is known as getting a **mortgage**.

The amount you can borrow depends on your wages. Banks and building societies do not usually lend the full amount that a property costs. If they do it is known as a **100% mortgage**. They usually lend up to 90% of the value of the property, depending on your salary. The difference between the amount you borrow and what the property costs is called the **deposit**. You have to supply this yourself!

Example **Calculating the deposit**

Mike and Jeda want to buy a flat and visit the local Estate Agents and check the local paper every night to find a suitable flat. Here is one flat they like the sound of.

CENTRAL FLAT

15 Roundabout Rd (Clarkson) 2 bedrooms, fitted kitcken, bathroom with coloured suite, lounge. Recently rewired. Modern plumbing. Good decorative order. RV £624. Offers over £21 000 to Rangers Estate Agents, Ibrox

The building society value the flat at £21 500 and will lend Mike and Jeda 90% of the valuation. They decide to offer £22 000 for the flat. How much deposit will they have to find?

Mortgage 90% of £21 500 = £19 350
Deposit £22 000 − £19 350 = £2650

Exercise 3 Look at each of these house adverts. The valuation and the price the buyer decides to offer are both given in each case. Calculate the deposit needed if the buyer is offered a mortgage of 90% of the valuation.

1 **NEW TOWN**
11 Union Street

This spacious well proportioned accommodation comprises: entrance porch, lounge, diningroom, 2 bedrooms, second hall/breakfast area, kitchen, bathroom and wc apartment. Extras included.
Rateable value, £653.
Viewing strictly by appointment
Tel 225 3387
Offers over £40,000 to:

DRUMMOND & CO
Tel 225 4487

Offer £41 000
Valuation £39 000

2 **STOCKBRIDGE**
18 Raeburn Place

OFFERS OVER £19,000
Most attractive modernised and newly decorated top-floor Flat (Fotheringham) close to all amenities. Accommodation comprises lounge, kitchen, double bedroom, shower-room and wc.
Rateable value, £218.

Russel & Aitken

SOLICITORS & ESTATE AGENTS
TEL: 225 5452

Offer £19 250
Valuation £19 000

3 **STOCKBRIDGE**
2 Leslie Place

Second-Floor Flat (Rostron) Lounge, kitchen / diningroom, double bedroom, boxroom and bathroom. Full gas central heating. Fitted carpets and gas cooker included. Entryphone. Stair being upgraded at seller's expense.
Rateable value, £548.
Further particulars from and offers over £28,000 to:
AITKEN KINNEAR & CO. WS.
QUEEN STREET
Tel 225 2914

Offer £33 250
Valuation £29 250

Information ⇔ A building society will arrange to have a flat surveyed before finalising a mortgage. There are 2 reasons for this. First they have to make sure that the property is safe and meets all the necessary building standards, and secondly to value the property. You have to pay for the survey. The costs are shown below.

Purchase price not exceeding:	£12 000	£15 000	£20 000	£25 000	£30 000	£40 000
Survey fee:	£25	£30	£35	£40	£45	£50

Add £5 for each additional £10 000 of value over £40 000.
These prices are exclusive of VAT. (VAT is usually 15%).

Example   **Survey fees**
Mike and Jeda's flat was valued at £21 500. How much will their survey bill be?

£21 500 is more than £20 000 but does not exceed £25 000
Survey fee £40 + 15% of £40 = £46

Exercise 4   Find the cost of having each of the properties in questions **1-4** surveyed. You will need to continue the table to answer some of these questions. Set out your working as shown in the above example.

1   A flat valued at £32 000.   2   A flat valued at £27 000.

3   A house valued at £39 000.   4   A house valued at £63 250.

Information ⇔ Once you have bought your flat you will start paying the mortgage. The table below shows monthly repayments for a loan of £1000 over different terms. If you pay income tax you get tax relief on your mortgage repayments.

	Interest rate 11%			
Repayment term in years	20	25	30	35
Monthly repayment per £1000 mortgage	£10.47	£9.90	£9.59	£9.42

Example   **Mortgage repayments**
Mike and Jeda have a mortgage of £19 350 over 20 years. How much will they repay each month?

Mortgage £1000 over 20 years   Monthly repayment £10.47
Mortgage £19 350 over 20 years   Monthly repayment £10.47 × 19.35 = £202.59

Exercise 5   Use the table above to calculate the monthly repayments for the 20 year mortgages given in questions 1 to 4.

1   £18 000   2   £24 000   3   £35 000   4   £36 000

5   Mary has just bought a new flat with a 25 year mortgage of £42 250. How much will she repay each month?

Rates

Information ⇨ Every householder pays **rates** to the local council. This money is used to pay for local services such as Education, Police, Fire Brigade, Parks, Street Lights, Cleansing, etc. The amount a householder pays depends on the **rateable value** of their house.

The council sets a **rate in the pound** which each householder is charged. The rates are then calculated from this word formula:

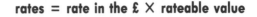

rates = rate in the £ × rateable value

Example

Calculating annual rates

Mike and Jeda's flat has a rateable value of £624. This information was given in the advert. The council have set the rate at 84p in the £. How much will Mike and Jeda have to pay in rates?

Rates = rate in the £ × rateable value
£0.84 × 624 = £524.16
Mike and Jeda's rates will be £524.16 per year.

Exercise 6

The Council rate for questions 1 to 6 is set at 84p in the £. Calculate the rates due on properties with rateable values of

1 £700	2 £535	3 £386
4 £620	5 £445	6 £800

7 Due to changes in the council's budget, the rate per £ drops to 82p. How much less will each of the householders in questions 1 to 6 pay in rates.

8 Raschid and Sue have had the rateable value of their flat re-assessed. The old rateable value was £750 and the new value is £825. Find the difference in their rates if the rate in the pound remains at 82p.

Information ➥ You may wonder how the council arrive at the figure for the rate per £. It is quite a simple calculation. The council knows the **total rateable value** for the area. They decide on a budget for the year and allocate money to each of the services they provide. This is called their **total estimated expenditure**. The rate in the £ is calculated from the word formula:

rate in the £ = total estimated expenditure ÷ total rateable value

Example **Calculating the rate in the £**
A council has a total estimated expenditure of £48 000 000 and a total rateable value of £64 000 000. What rate in the £ must be set?

Rate in the £ = total estimated expenditure ÷ total rateable value
£48 000 000÷64 000 000 = £0.75

Information ➥ If the rate in the £ does not work out exactly, round up to the next 1p. The extra money collected is called a **surplus**.

Example **Calculating the rate in the £ with a surplus**
The following year the council estimates its total expenditure to be £51 000 000. What rate in the £ should they set if their total rateable value is still £64 000 000?

Rate in the £ = total estimated expenditure ÷ total rateable value
£51 000 000÷64 000 000 = £0.7969 to 4 decimal places

If the council charged 79p in the £ they would collect £0.79×64 000 000 = £50 560 000, not the £51 000 000 needed.
If the council charged 80p in the £ they would collect £0.80×64 000 000 = £51 200 000, a surplus of £200 000.

Information ➥ If the rate in the £ does not work out exactly, round up to the next 1p. The extra money collected is called a **surplus**.

Exercise 7 For questions **1** to **6** calculate the rate in the £ and, where appropriate, the surplus.

	Total estimated expenditure	Total rateable value
1	£30 000 000	£40 000 000
2	£81 600 000	£96 000 000
3	£41 400 000	£36 000 000
4	£47 000 000	£53 000 000
5	£96 000 000	£85 000 000
6	£79 000 000	£95 000 000

7 A local council has a total rateable value of £37 000 000. Their total estimated expenditure for the following year is £30 340 000. What rate in the £ must they set?

8 A council estimate their years expenditure to be £97 000 000. Their total rateable value is £85 000 000. What rate in the £ must they set and what surplus will they collect?

Insurance

Information ➪ When you buy a house or flat your building society or bank will insure the building on your behalf. The amount the building will be insured for will be much higher than the purchase valuation. It costs a lot more to rebuild a house! This is done to protect their investment should the property be destroyed or damaged. You will be charged the annual **premium**. The cost of property insurance depends on its value and the estimated cost of rebuilding it.

Example

Calculating property insurance
The Hawk Insurance Company charge £1.62 per £1000 for a property to be insured. Mike and Jeda's flat is valued at £47 000. How much will they be charged for insurance?

Value: £1000   Insurance £1.62
Value: £47 000   Insurance £1.62 × 47 = £76.15

Exercise 8

Calculate how much it would cost to insure the properties in questions **1-15** with Hawk Insurance Company.

1 A semi-detached house valued at £58 000.

2 A flat valued at £35 000. **3** A house valued at £42 000

4 A terraced house valued at £48 000. **5** A flat valued at £23 000

6 A flat valued at £43 000. **7** A house valued at £75 000

8 A villa valued at £70 000. **9** A flat valued at £43 500

10 A house valued at £44 550. **11** A villa valued at £38 500

12 A detached house valued at £61 650. **13** A cottage valued at £68 250

14 A flat valued at £31 775. **15** A garage valued at £3450

Information ➫ It also makes sense to insure the contents of your house against damage or theft. There are 2 main types of contents insurance. These are **Replacement** and **New for Old**.

Replacement insurance pays for your losses but allows for depreciation. This means that you won't get as much as you paid for most of your belongings.

New for Old insurance replaces your lost goods with equivalent new ones. This means that if your television is stolen 3 years after you bought it and you have a replacement insurance policy on your contents, you will get the price you paid for it less 3 years depreciation. Insurance companies use standard depreciation rates for most items.

If you have a New for Old policy then you will get the price of a new television.

Example

Calculating contents insurance premiums
The Hawk Insurance company will insure contents at the following rates:
  Replacement: 42.6p per £100 insured
  New for Old: 48.3p per £100 insured
Mike and Jeda estimate that the contents of their flat are worth £9200. How much will it cost to insure their contents under each scheme?

Replacement
Value: £100   Insurance: £0.426
Value: £9200   Insurance: £0.426 × 92 = £39.19

New for Old
Value: £100   Insurance: £0.483
Value: £9200   Insurance: £0.483 × 92 = £44.44

Exercise 9   Calculate the cost of insuring the contents valued below under each of the 2 schemes.

1 £12 000	2 £9800	3 £15 000	4 £6500
5 £10 000	6 £23 000	7 £8550	8 £11 560

9   Jemma has moved into a new house and wants to insure both the buildings and the contents with Hawk. The building is valued at £62 500 and the contents at £12 350. How much will she pay altogether for insurance if she chooses to insure the contents on the basis of New for Old?

Information ↩ As you have seen, before you can insure the contents of your house you must have a good idea of their total value. The form below shows one way to work out the total value.

Investigations

1   Use mail-order catalogues or look in shops to find the price of each item listed here. Copy this form and fill in the cost of each item. Calculate the total value of the contents.

Calculate the cost of insuring the contents.

```
LOUNGE                              BEDROOM 1
Carpet .............................   Carpet ...........................
Curtains ...........................   Curtains .........................
Suite...............................   Bed ..............................
T.V.................................   Wardrobes.........................
Video...............................   Lights............................
Fire ...............................   Clothes ..........................
Other...............................   Other.............................
TOTAL ..............................   TOTAL ............................

                                       BEDROOM 2
                                       Carpet ...........................
                                       Curtains .........................
KITCHEN                                Bed ..............................
Units...............................   Clothes ..........................
Utensils............................   Other.............................
Flooring ...........................   TOTAL ............................
Other...............................
       .............................   OTHER GOODS
TOTAL ..............................   Sports equipment..................
                                       Camera............................
                                       Other.............................


TOTAL TO BE INSURED £.....................
```

2   Use a form similar to the one shown above and list the contents of your own house. Use a mail-order catalogue to find the price of each item. Remember to do all the rooms. Calculate the total value of the contents. Is your insurance adequate?

3   Use the Hawke Insurance Company rates to calculate: the cost of insuring the contents of
    (a) the house in investigation 1, above.
    (b) your own house
    Calculate the premium using both replacement and New for Old insurance.

    Which type of insurance cover would you choose? Give your reasons.

Life assurance

Information ⇨ Insurance is taken out in case something happens. **Assurance** is taken out when you know something is going to happen. It is common for people to take out assurance on their life. (They know they will die some time.) When they die their dependants (close relatives) will receive a payment from the assurance company.

Many companies try to sell different types of personal assurance. You can take **Whole Life Policies** which pay out when you die and **Endowment Policies** which pay out after a set time or sooner if you die before this time. The tables below give the monthly costs.

Monthly premiums per £1000 assured									
WHOLE LIFE		**ENDOWMENT**							
Age next birthday		Age next birthday	Term						
			15 Years	20 Years	25 Years	30 Years	To age 55	To age 60	To age 65
	£		£	£	£	£	£	£	£
20	0.60	20	4.35	2.98	2.17	1.65	1.30	1.05	0.88
21	0.62	21	4.35	2.98	2.17	1.66	1.37	1.10	0.92
22	0.64	22	4.35	2.98	2.17	1.66	1.43	1.15	0.96
23	0.67	23	4.36	2.98	2.18	1.66	1.51	1.21	1.00
24	0.69	24	4.36	2.98	2.18	1.67	1.58	1.26	1.04
25	0.72	25	4.36	2.98	2.18	1.67	1.67	**1.33**	1.09
26	0.75	26	4.36	2.98	2.18	1.67	1.76	1.39	1.14
27	0.78	27	4.36	2.99	2.19	1.68	1.86	1.46	1.19
28	0.81	28	4.36	2.99	2.19	1.69	1.96	1.54	1.25
29	0.84	29	4.36	2.99	2.20	1.69	2.08	1.62	1.31
30	0.87	30	3.36	2.99	2.20	1.70	2.20	1.70	1.37
31	0.91	31	4.37	3.00	2.21	1.71	2.34	1.80	1.44

Example **Calculating assurance premiums**
Jemma decides to take out an endowment policy. She is 24 and wants the policy to mature when she is 60. She wants a policy for £5500. How much will she pay per month?

Value: £1000   Premium: £1.33 (from table, across: 'Age 25', down: 'To age 60')
Value: £5500   Premium: £1.33×5.5 = £7.32

Exercise 10 Copy and complete the table below.

	Name	Age next birthday	Policy	Term	Value	Premium
1	A Akinhead	25	Whole Life	—	£10 000	
2	C Straight	30	Endowment	25 yrs	£6000	
3	L Driver	28	Endowment	to age 60	£8000	
4	P Nightly	29	Whole Life	—	£7500	
5	E Tone	31	Endowment	30 yrs	£6200	
6	T Break	24	Endowment	to age 65	£9400	

7 What happens to the monthly cost of assurance if you take it out as you get older? Explain your answer.

Flat-mates

Sue and Angela are unsure about buying or renting a flat. They try to list the advantages and disadvantages of each.

Buying a flat

For:
A good investment.
A place of our own.
Income tax relief on the mortgage repayments.
No landlords to throw you out.
Money not 'wasted' on renting.

Against:
Expensive to buy.
A lot of bills to pay in the buying process.
Responsiblity for repairs.

Renting a flat

For:
Landlord responsible for repairs.
Don't have to furnish the flat.

Against:
Landlord can give notice to leave.
Renting a flat means you never get to own it.
No income tax relief on rent.

Can you think of any other advantages or disadvantages for renting and buying? Write them down under the appropriate heading.

Sue earns £8600 per year. Angela earns £545 per month which is £6540 per year. These earnings mean that the two girls could probably get a joint mortgage for 2½ times Sue's annual salary plus 1½ times Angela's annual salary. (Sue's salary is the higher of the two.)

$$2\frac{1}{2} \times £8600 = £21\ 500.$$
$$1\frac{1}{2} \times £6540 = £\ 9\ 810.$$
Total mortgage = £31 310

Sue and Angela could get a mortgage for £30 000 over 20 years in which case their joint monthly payments would be (from table on p.212)

£1000   over 20 years Monthly repayments £10.47
£30 000 over 20 years Monthly repayment £10.47 × 30 = £314.10

Remember they will also have to pay for insurance, rates, furniture, bills . . . Do you think they can afford to buy a flat?

Investigations   1   Note all the food you eat in 1 week. Go to the supermarket and find the total cost of the food.

What is your personal weekly food bill? Will this vary much from week to week?

2   Most homes use electricity or gas (or both) to provide heating, lighting, and cooking facilities. The electricity and gas boards then send out bills every 3 months.

Find how much has been paid in your home for electricity and gas bills over the last year. Draw a graph of the bills.

What do you notice?

Which months were cheapest?

Why do you think this happens?

How much should you budget each month for fuel?

3   You are thinking of buying a flat. What is the current mortgage interest rate? How much will a solicitor or estate agent charge for their services? List all the things you have to arrange before buying the flat.

4   Look at the house adverts in your local paper or visit a local estate agent. Choose a house or flat that you like and work out

(a) the cost of a survey

(b) the mortgage you would need to buy it, assuming that the building society give a 90% mortgage

(c) the monthly repayments for the mortgage

(d) the cost of insuring the buildings.

5   You have just bought a 1 bedroom flat and have saved up £1500 to furnish it. List your priorities for furniture and equipment. Remember you have only £1500 to spend. Once you have decided what you can buy, compare your list and prices with others in your class.

13 Trigonometry problems

Do you remember? ⇐

In this unit it is assumed that you know how to:

1 calculate sides and angles in right-angled triangles

2 show the direction of a 3-figure bearing on a sketch.

Exercise 0 will help you check. Can you answer all the questions?

Exercise 0

Sides and angles

Remember **SOH CAH TOA?** $\sin = \dfrac{\text{opp}}{\text{hyp}}$ $\cos = \dfrac{\text{adj}}{\text{hyp}}$ $\tan = \dfrac{\text{opp}}{\text{adj}}$

Use sin, cos or tan ratios to find the size of angle x in the following right-angled triangles.

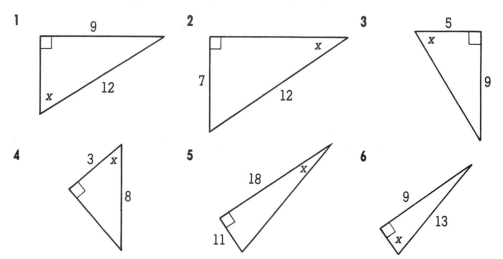

Use sin, cos or tan ratios to find the length of side s in the following right-angled triangles.

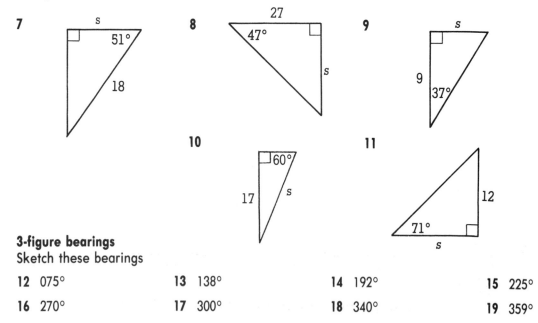

3-figure bearings
Sketch these bearings

12 075°	**13** 138°	**14** 192°	**15** 225°
16 270°	**17** 300°	**18** 340°	**19** 359°

Hanging around

Dawn is entering a hang-gliding contest.

One part of the competition is an accuracy dive. This event means that Dawn will have to hang-glide from the top of a hill, in a direct line, down to a target area. Fortunately there is no wind on the day of the competition so she glides in a straight line.

The diagram shows the competition area. What will Dawn's angle of descent have to be if she is to hit the target?

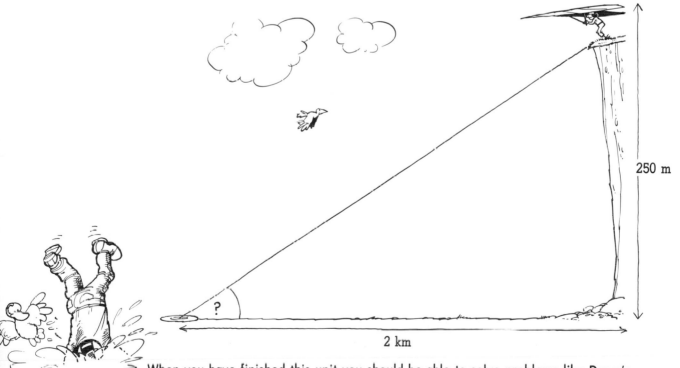

250 m

2 km

When you have finished this unit you should be able to solve problems like Dawn's.

Gradient

Information **Gradient** is a measure of steepness.
In road signs gradient is shown either as
a ratio or as a percentage.

Example **Changing gradients to percentages**
What does a gradient of 1 in 4 mean?
What would this be as a percentage?

A gradient of 1 in 4 means that for
every 4 metres measured along the road
you go down (or up) 1 metre vertically.
As a percentage this would be

$$1 \text{ in } 4 = \frac{1}{4} = 25\%$$

Example **Calculating the size of an angle**

This road sign shows the gradient of the
road to be 1 in 10. This means that for
every 10 m measured along the road
you go down 1 m vertically.

At what angle to the horizontal is the
road if the gradient is 1 in 10?

Make a sketch and mark in all the
known information.
Mark the angle you want as x.
Label the sides **h**, **o** and **a**.
You need to use sides **o** and **h** so use
sin (SOH)

$$\sin x° = 1 \div 10 = 0.1$$
$$x = 5.7°$$

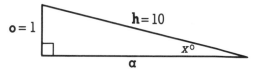

Example **Calculating a length**
The gradient of a road is 20%. What height does the road rise vertically over a distance
of 1.5 km?

This problem can be solved by using a scale factor and similar triangles.

20% = 20 ÷ 100 so for every 100 m measured along the road you rise 20 m vertically.
From similar triangles scale factor = 1500 ÷ 100 = 15

The road rises 20 × 15 = 300 m

Exercise 1   The following problems can be solved by using either the sine ratio or similar triangles. Set out your working as shown in the examples.

1   The steepest classified road in Britain has a gradient of 1 in 4. What angle is this to the horizontal?

2   Satan's slide is the name given to a stretch of the A93 road near Glenshee in Scotland. It reaches an altitude of 2200 feet (670 metres) and is the highest classified road in Britain.

   The last section of the road climbs at a gradient of 12% for a distance of 400 metres. What height does this part of the road climb?

3   A new chairlift is designed to climb a height of 320 metres. The designer calculates that the chairs will climb at an angle of 37° to the horizontal. What length of cable will be needed? Remember that the cable must be long enough to go up and down the hill. Allow an additional 8 m of cable to fit round the wheels at the top and bottom pylon.

4   The simplified sketch below shows the plan for a new chairlift. What is the total height that the skiers will rise when using the lift?

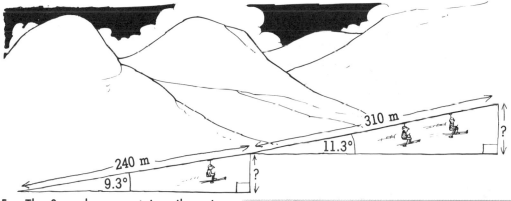

310 m
11.3°
?
240 m
9.3°
?

5   The Snowdon mountain railway is 7 km long and climbs a height of 980 m. What is the average angle at which the train climbs?

6   A ladder 2.5 m long is resting against a wall. How high up the wall does the ladder reach if it is leaning at an angle of 68° to the horizontal?

7   A ramp 5.4 m long is used on a building site to run wheelbarrows up and down. The top of the ramp is 1.7 m higher than the foot. What angle does the ramp make with the horizontal?

Angles of elevation and depression

Information ➪

An **angle of elevation** is measured from the horizontal in an upward direction.

An angle of **depression** is measured from the horizontal in a downward direction.

You have already solved problems involving angles of elevation and depression by scale drawing. Now you are going to use trigonometry.

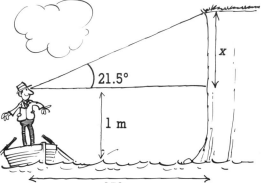

Example

Angle of elevation
The sketch shows a boat 150 metres away from a cliff. The angle of elevation from the boat to the top of the cliff is 21.5°. How high is the cliff?

Sketch the triangle
Mark in the information
Label the height x
Label the sides **h**, **o** and **a**
Need **o** and have **a** so use tan (TOA)

$\tan 21.5° = x \div 150$
$0.394 = x \div 150$
$x = 0.394 \times 150$
$x = 59.1$ m

The angle of elevation is measured 1 m above sea level so the cliff is 60.1 m high.

Exercise 2

These problems can be solved by using the tangent ratio. Set out your working as shown in the example.

1   The sketch shows a boat 250 metres away from a cliff. The angle of elevation from the boat to the top of the cliff is 18.7°. How high is the cliff?

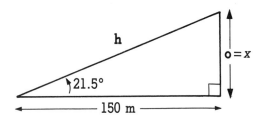

2 From the top of a cliff 40 m high, Paul sights a fishing boat and measures the angle of depression to the bottom of the boat to be 28.2°. How far is the boat from the foot of the cliff?

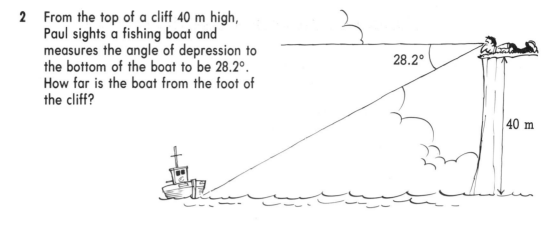

3 When asked to guess the height of the local church spire, Mary made the following measurements. At a distance of 70 metres the angle of elevation is 14.3°. How high is the church spire?

4 From his bedroom window Robert measured the angle of depression to the end of the garden to be 14.6°. If his bedroom window is 6 metres above the ground how long is the garden?

5 Find the angle of elevation to the top of a building 18 metres high from a point 30 metres from the foot of the building.

6 Find the angle of depression from the top of a tower 35 metres high to a point 55 metres from the foot of the tower.

7 The sketch shows a tower 50 metres high and 2 points A and B which are in line with the base of the tower.

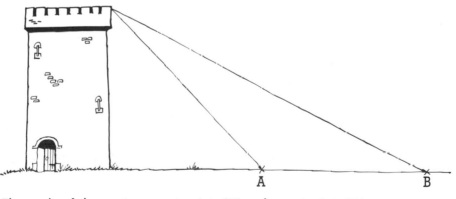

The angle of depression to point A is 37° and to point B is 26°.
(a) How far is point A from the foot of the tower?
(b) How far is point B from the foot of the tower?
(c) Calculate the distance between points A and B.

8 Assad is standing on a cliff top 46 metres high in line with 2 buoys. The angles of depression to the buoys are 23.6° and 13.8°. Calculate the distance between the 2 buoys.

9 Norman is on the roof of his house which is 7.4 metres high. He is 63 metres away from a block of flats and measures the angle of elevation to be 18.3°. How high are the flats?

Bearings and trigonometry

Information ↝ Trigonometry is used to help in navigation. By using right-angled triangles you can change any given distance and bearing into 2 distances. One distance is calculated as either North or South while the other is calculated to be either East or West. This helps to plot positions on maps or charts and is called **resolving a bearing**.

Examples

Resolving a bearing

1 A ship leaves port and sails on a bearing of 073° for a distance of 35 km. How far is the ship North and East of the port?

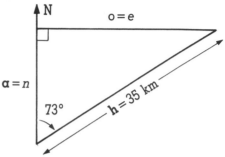

(a) Make a sketch of the route.
(b) Complete the right-angled triangle.
(c) Mark in the known information.
(d) Mark the sides you want *n* and *e*.
(e) Label the sides **h**, **o** and **a**.

To find the distance North (*n*)
a is wanted and **h** is known so use cos (CAH)

$$\cos 73° = \frac{n}{35}$$

$$0.292 = \frac{n}{35}$$

$$n = 0.292 \times 35$$
$$n = 10.2 \text{ km}$$

To find the distance East (*e*)
o is wanted and **h** is known so use sin (SOH)

$$\sin 73° = \frac{e}{35}$$

$$0.956 = \frac{e}{35}$$

$$e = 0.965 \times 35$$
$$e = 33.8 \text{ km}$$

The ship is 10.2 km North and 33.8 km East of port.

2 A plane flies for 240 km on a bearing of 228°. How far West has it flown?

(a) Make a sketch of the route.
(b) Complete the right-angled triangle.
(c) Mark in the known information.
(d) Mark the side you want *w*.
(e) Label the sides **h**, **o** and **a**.
  a is wanted and **h** is known so use cos (CAH).

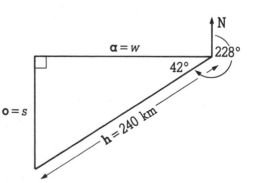

$$\cos 42° = \frac{w}{240}$$

$$0.743 = \frac{w}{240}$$

$$w = 0.743 \times 240$$
$$w = 178 \text{ km}$$

The plane has flown 208 km West.

Exercise 3

1 A plane flies 160 km on a bearing of 057° from a point P. How far North and East is the plane from P?

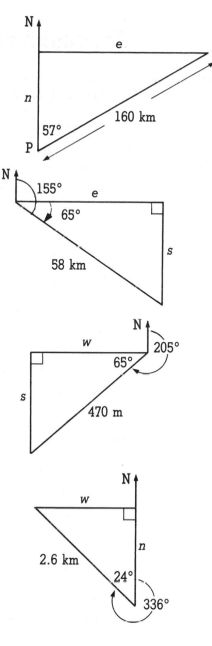

2 The SS Uganda sails from port on a bearing of 155° for a distance of 58 km. How far South and East has the ship sailed?

3 Said goes orienteering and runs 470 m on a bearing of 205°. How far South and West is he from his starting point?

4 Susan, while out hillwalking, walks a distance of 2.6 kilometres on a bearing of 336°. How far North and West is she from her starting point?

5 A light aircraft takes off and flies for 69 kilometres on a bearing of 138°. How far South of the airport has the plane flown?

6 A helicopter takes off from an oilrig and flies 220 kilometres on a bearing of 257° to Aberdeen. How far is Aberdeen West of the oilrig?

7 A ship sails from P 40 kilometres East to Q then 27 kilometres North to R.
 (a) Sketch the journey.
 (b) Calculate the size of angle QRP.
 (c) What bearing must the Captain steer to return to P?

8 A helicopter takes off from base and flies for 85 kilometres on a bearing of 270°. It then turns and flies for 36 kilometres on a bearing of 180°.
 (a) Make a sketch of the journey.
 (b) What bearing must be followed to return to base?
 (c) How far is it back to base?

Trigonometry mixtures

Information ➷ This exercise consists of a number of questions which can be solved by using the techniques you have studied in this unit.

Example

Using trigonometry
As part of a Navy display, sailors demonstrate moves on the yard arms and mast.

The first yard arm is 5 m above the deck, and the rope joining the deck to the end of the yard arm is 8 m long.

What angle does the rope make with the yard arm?

Make a sketch, marking in all the information. Label the angle you want $x°$.
Label the sides **o**, **h**, and **a**.
Know **o** and **h**, so use sin (SOH).

$$\sin x° = \frac{o}{h}$$

$$= \frac{5}{8}$$

$$\sin x° = 0.625$$

$$x° = 38.7°$$

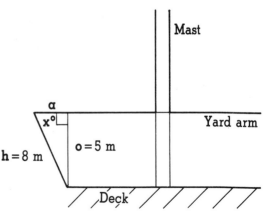

Exercise 4

Questions 1 and 2 refer to the sketch below.

1 A flag pole is supported by wire guys. Find the angle each guy makes with the ground if each guy is 4 metres long and is attached to the flag pole at a height of 2.2 metres.

2 How far from the foot of the flag pole are the guys anchored to the ground?

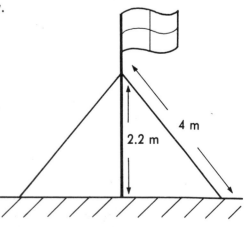

3 A spotlight is situated on the ground and angled to illuminate the top of a tower 28 metres high. The light is 15 metres from the foot of the tower. At what angle must the spotlight be fixed?

4 A ladder 2.8 metres long rests against a wall. The foot of the ladder makes an angle of 64° with the ground. How high up the wall will the ladder reach?

5 The sketch below shows a death slide set up from a cliff. The cliff is 32 metres high and the rope is set at an angle of 30° to the horizontal.

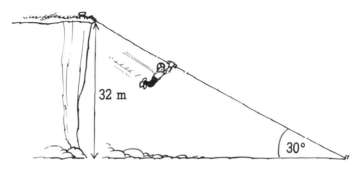

(a) How far from the base of the cliff is the rope anchored?
(b) How long is the rope?

6 The sketch shows 2 spotlights arranged to light up the same spot of stage, midway between the 2 lights. The spotlights are 6 metres apart and 9 metres above the stage. At what angle from the vertical are the spotlights fixed?

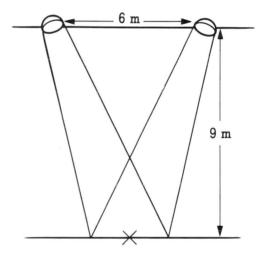

7 The sketch shows a chute used to send goods between 2 floors in a warehouse.

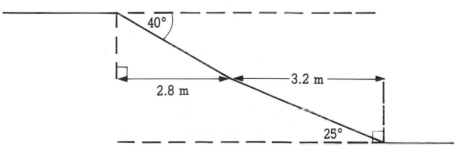

(a) What is the total length of the chute?
(b) What is the height difference between the floors?

8 A stunt kite is designed to fly at its maximum height when the control lines are angled at 62° to the horizontal. What is the maximum height the kite can reach if the control lines are 100 ft long?

9 To ensure easy drainage a waste pipe must have a minimum gradient of 1 in 40. A pipe is 30 m long. What is the height difference between its two ends?

10 The sketch opposite shows a garden hut.

A new roof is to be put on the hut.
(a) Calculate the length of the sloping side of the roof.
(b) Calculate the area of wood needed for each side of the roof.

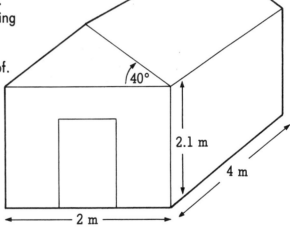

11 The photograph shows an unclimbable tree which is to be cut down. It is a bright sunny day and Dave is trying to find the height of the tree. All he has is a 10 metre tape and a metre stick. Dave measures the length of the tree's shadow. Next he measures the length of the shadow cast when he holds the metre stick vertically.

Explain how he can use these measurements to find the height of the tree.

Hanging around

Dawn wanted to work out what her angle of descent would need to be to hit the target in the hang-gliding competition.

Here is a sketch she drew, marking in all the information, and labelling the sides **h**, **o** and **a**.

Dawn labelled the angle x°
Know **o** and **a** so use tan.

$$\tan x° = \frac{2000}{250} = 8.000$$

$$x° = 82.9°$$
(from tables or calculator)

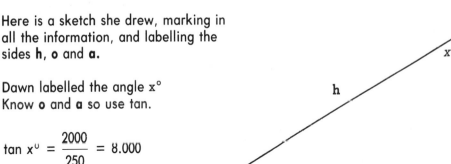

The angle of descent is the difference between this angle and the horizontal.

So the angle of descent is 90° − 82.9° which is 7.1°.

Investigation   **Designing stairs**
When architects design buildings they have to make sure that they conform to the current building regulations.

Here are the regulations concerning stairs. (The diagram shows what architects call the rise (r) and the going (g).)

1   The steps must not be steeper than 38° to the horizontal.

2   2r+g must be between 550 mm and 700 mm where
    r is the rise measured in millimetres
    g is the going measured in millimetres.

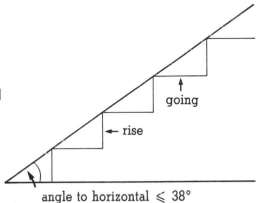

3   The minimum measurement for the going is 250 mm.

4   No more than 16 steps are allowed in 1 flight.

Design a flight of stairs to join 2 floors which are 3 metres apart.

How steep are your stairs?

14 Maths on the move

Do you remember? ⬅

In this unit it is assumed that you know how to:

1   multiply and divide
2   change hours and minutes to minutes
3   change a fraction of an hour to minutes
4   change 12-hour to 24-hour time.

Exercise 0 will help you check. Can you answer all the questions?

Exercise 0

Multiplying and dividing

1 28×47	2 146×34	3 1296×65
4 34.75×78	5 810.62×45	6 7.13×70
7 $650 \div 50$	8 $345 \div 60$	9 $230 \div 40$
10 $125 \div 4.5$	11 $25.7 \div 8.4$	12 $125.75 \div 6.5$

Converting hours and minutes to minutes.

How many minutes are there in:

13 1 hour 30 mins	14 2 hours 20 mins	15 2 hours 45 mins
16 4 hours 17 mins	17 2 hours 37 mins	18 3 hours 29 mins
19 1 hour 18 mins	20 5 hours 36 mins	21 3 hours 57 mins

Rewrite these minutes in hours and minutes.

22 75 mins	23 83 mins	24 95 mins
25 123 mins	26 154 mins	27 180 mins
28 247 mins	29 310 mins	30 348 mins

Converting fractions of an hour to minutes

Write these fractions of an hour in minutes or hours and minutes.

31 0.1	32 0.2	33 0.3	34 0.4
35 1.5	36 0.6	37 1.7	38 1.9

Converting 12-hour to 24-hour time

Rewrite these 12-hour times in the 24-hour time.

39 8 am	40 11:30 am	41 12 midday
42 2:45 pm	43 8:20 pm	44 11:55 pm

45 Quarter past two in the afternoon.   46 Quarter to ten at night.

Youth club outing

Information ➾ Mary and Kirsten want to hire a minibus for their youth club outing to the Lake District, which will last a whole day.

They ask 2 local firms for their rates:

Which should they hire for the outing?

After working through this unit you should be able to help Mary and Kirsten choose the cheaper company.

Timetables

Information ⌫ The railway network in Great Britain is quite large. In order to make the best use of railways, and for convenience to passengers, each train service runs to a **timetable**. Here is a copy of the Aberdeen to London train timetable. (*d* means depart, *a* means arrive.)

SOUTHBOUND (MONDAYS TO SATURDAYS) DAYTIME SERVICES 30th SEPTEMBER, 1985 to 10th MAY, 1986						A		B	C			D		
Aberdeen	d	—	0600	—	0800	—		1000	—	—	—	1400	—	1600
Stonehaven	d	—	0616	—	—	—		1016	—	—	—	1416	—	1616
Montrose	d	—	0637	—	0837	—		1037	—	—	—	1437	—	1637
Arbroath	d	—	—	—	0851	—		1051	—	—	—	1451	—	1651
Dundee	d	—	0712	—	0912	—		1112	—	—	—	1512	—	1712
Leuchars	d	—	—	—	—	—		1124	—	—	—	—	—	1724
Kirkcaldy	d	—	0747	—	0947	—		1147	—	—	—	1547	—	1747
Inverkeithing	d	—	0804	—	—	—		—	—	—	—	1604	—	1804
Edinburgh	d	0735	0835	0935	1035	1055	1135	1235	1335	1435	1535	1635	**1735**	1835
Dunbar	d	—	0856	—	—	1116	—	—	—	—	—	—	1756	—
Berwick-upon-Tweed	d	0817	—	1017	—	1142	—	1317	—	1517	—	1717	1820	—
Newcastle	a	0912	1016	1112	1210	1238	1310	1412	1510	1612	1710	1812	1915	2010
Durham	a	—	—	—	—	1253	—	—	—	—	—	—	1931	2025
Darlington	a	—	1046	1142	—	1310	1340	1442	1540	1642	1740	1842	1948	2040
York	a	1010	—	1214	—	1342	1412	—	1612	1714	1812	1914	2020	2112
Doncaster	a	1035	—	1239	—	1407	—	—	—	—	—	1939	2045	2136
Peterborough	a	—	1229	—	—	1503	—	1626	—	1829	—	2039	2157	2232
London King's X	a	1219	1325	1423	1508	1603	1616	1723	1817	1923	2019	2140	**2304**	2331

A = The Flying Scotsman   B = The Highland Chieftain   C = The Aberdonian   D = The Talisman

Example

Using a timetable
Angela is going to stay with friends in London for the weekend. She wants to leave Edinburgh, where she lives, after work. She can be at the station by 5:15 pm. What train can she catch? How long will the journey to London take?

5:15 pm is 1715 in the 24 hour clock.
Looking at the timetable you can see that the first train after 1715 leaves at 1735 and arrives in London at 2304. To find the journey time we subtract the departure time from the arrival time.

hr	min		hr	min
23	04	You cannot subtract 35 from 04 so	22	64
− 17	35	rewrite 23 hours 04 minutes	− 17	35
		as 22 hours 64 minutes	5	29

The journey will take 5 hours 29 minutes.

Exercise 1 Use the Aberdeen to London train timetable to answer these questions.

1 When is the first train from Edinburgh to London each day?

2 When does this train arrive in London?

3 When does the first train leave Aberdeen each day for London? How long does the journey take?

4 Helen wants to travel from Edinburgh to York, leaving Edinburgh in the early afternoon. When is the first train she can catch? How long does the train take to travel to York?

5 Philip must be in York by 3 pm to attend a meeting. He is travelling from Newcastle. When does the latest train he can catch leave Newcastle? How long does the journey take?

6 Work out the journey times for all 5 trains from Aberdeen to London. Which is quickest?

7 Which is the quickest train between Edinburgh and London? Why do you think this train is quickest?

8 Darren wants to travel by train from Aberdeen to York. He is meeting his mum in York at 2 pm. What train should he catch? How long will he have to wait when he arrives in York before meeting his mum? What is the total time he will spend travelling?

9 Shirley wants to travel from her home in Leuchars to her sister's in Peterborough. She wants to travel during the day. Plan her journey. How long will she spend travelling?

10 Mehmet has to get from Newcastle to Darlington for a meeting at 3 pm. He estimates that it will take him 1 hour to get from the station in Darlington to the hotel in which the meeting is to be held. What train should he catch from Newcastle?

11 Linda lives in Kirkcaldy and is going to Doncaster to watch the racing. Plan her journey. What is the earliest that she can arrive in Doncaster? How long will she spend travelling?

12 Carol has an interview in London at 2:30 pm. She lives in Dunbar. Which train should she catch? How much time will she have in London before her interview?

On the road

Information ➣ If you want to drive somewhere - particularly longer trips — it is best to plan your route first. In Great Britain, road maps show all the Motorways and main roads. Each motorway (M) and main road (A) has a letter and number. For example M6 or A1. These are given on the road signs.

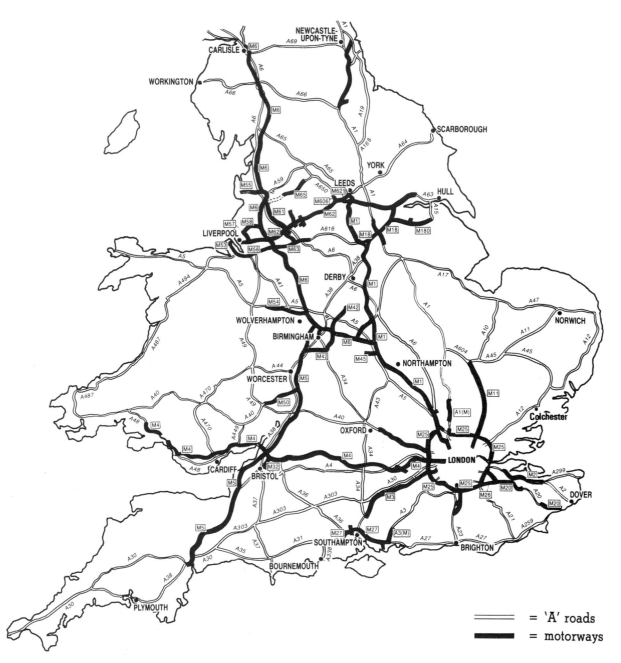

====== = 'A' roads

██████ = motorways

Examples   **Using a road map**
Paul has to travel from his house in Northampton to a conference in Cardiff, South Wales. Plan his route.

A43   Northampton to Oxford

A34   Oxford to M4

M4   to Cardiff

Michelle Foster lives in Carlisle and is considering driving down to see her brother who lives in Colchester, 50 miles North East of London. Use the road map to work out a route for Michelle.

Michelle's route is:
Leave Carlisle on the A6   A66   A1   M25   A12

Exercise 2

Use the road map on the previous page to plan a route for the following journeys.

1 Derby to Leeds
2 Workington to Newcastle-Upon-Tyne
3 Birmingham to York
4 Carlisle to Leeds
5 Birmingham to Hull
6 Cardiff to Norwich
7 Carlisle to Southampton
8 Bristol to Norwich
9 Liverpool to Dover
10 Hull to Plymouth

11 Manmohan wants to drive from Northampton to Southampton. Plan a route for him.

12 Hugh and Lorna are travelling from a friend's house in Derby to Hugh's parents in Oxford. Plan a route for them.

13 Pat and Cathy live in Liverpool. They are going to Bournmouth, by car, for a holiday. Plan a route for them.

14 Douglas, who lives in Scarborough, plays for an amateur football team. The team have to travel from Scarborough to a tournament in Wolverhampton. Plan a route for them.

15 Neil plays the cello. He has to travel from Worcester to Brighton to play at an evening concert in the Pavilion. Plan his route.

16 John lives in Derby. He wants to travel by car to Dover to catch a hovercraft. Plan his route.

Mileage charts

You can work out how far it is from one place to another by road by looking at a **mileage chart**. These are often given in road atlases and on road maps. All the main cities and towns are usually included. Here is an example of a mileage chart.

```
Birmingham
 85  Bristol
107   46  Cardiff
197  275  295  Carlisle
202  198  234  393  Dover
293  373  395   98  457  Edinburgh
291  372  393   97  490   45  Glasgow
136  227  246  155  278  229  245  Hull
453  537  558  260  623  158  175  393  Inverness
115  216  236  122  265  205  215   59  367  Leeds
 98  178  200  125  295  222  220  126  385   72  Liverpool
 88  167  188  118  283  218  214   97  379   43   34  Manchester
198  291  311   58  348  107  150  121  268   91  170  141  Newcastle
161  217  252  284  167  365  379  153  528  173  232  183  258  Norwich
339  420  443  144  507   43   60  277  115  251  269  263  153  413  Perth
199  125  164  391  290  488  486  341  851  328  294  281  410  236  535  Plymouth
128   75  123  339  155  437  436  253  599  235  241  227  319  192  483  155  Southampton
128  221  241  117  274  191  208   38  354   24  100   71   83  185  238  340  252  York
118  119  155  307   77  405  402  215  569  196  210  199  280  115  453  215   76  209  London
```

Example

Using a mileage chart

Jane and Robert are driving from their home in Hull to visit Robert's cousin in Southampton. Use the mileage chart above to work out how far they have to travel.

To find the distance between Hull and Southampton look down on the mileage chart from Hull and along from Southampton. The reading on the chart is 253. So it is 253 miles from Hull to Southampton.

Exercise 3

Use the mileage chart above to find the distance between the pairs of towns or cities in questions 1-6.

1 Edinburgh to Glasgow   2 Glasgow to Newcastle

3 Edinburgh to York   4 Carlisle to Leeds

5 Glasgow to Hull   6 Cardiff to Norwich

7 Which city is 136 miles from Birmingham?

8 Which city is 295 miles from Liverpool?

9 Ian and Janet are two sales reps for the same firm in Liverpool. One day Janet has to travel to York then to Hull. On the same day Ian has to travel to Leeds then Manchester. Both have to get back to Liverpool. Who travels the furthest and by how many miles?

Information ⇨ Here is a road map showing the distances between some towns in the North of Scotland. We can use the map to make a distance table. The towns are usually arranged in alphabetical order on a mileage chart.

Example

Constructing a mileage chart

Elgin to Fraserburgh   60 miles
Elgin to Inverness   22+15 = 37 miles
Elgin to Peterhead   60+18 = 78 miles
Elgin to Nairn   22 miles
Fraserburgh to Inverness 60+22+15 = 97 miles
and so on until the table is complete.

Exercise 4 Draw a mileage chart for each of the maps below. Where there is a choice of routes the shortest distance should be chosen.

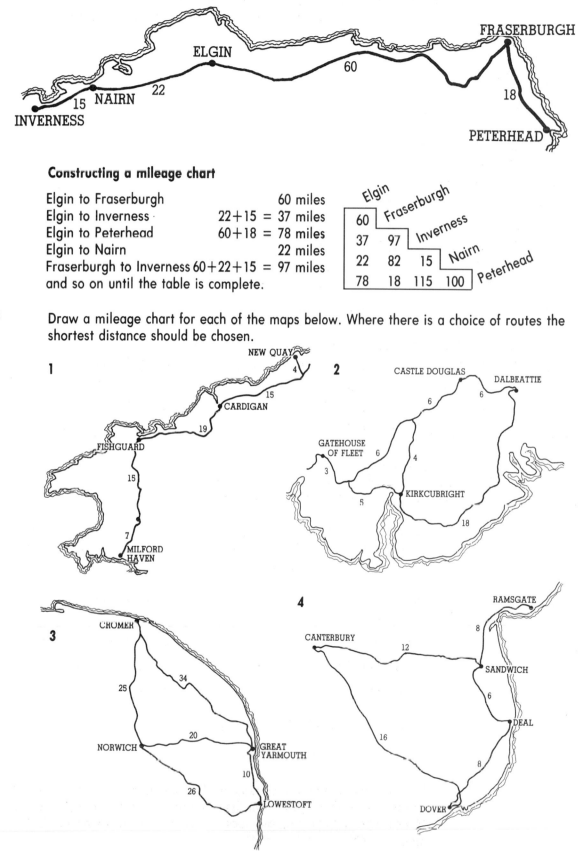

Motorway maps

Information➦ You may have travelled on a Motorway and noticed numbers on the road signs as you enter or leave the motorway. These are called **Junction Numbers**. Each motorway has a number of **junctions** and maps are made showing all the junction numbers, the towns reached by each junction and the distances between the junctions.

Example **Using a motorway map**
Here is the M25 motorway map. From the map you can find lots of information.

You would leave the M25 at Junction 28 for the A12.

From Junction 10 Guildford is 8 miles away on the A3.

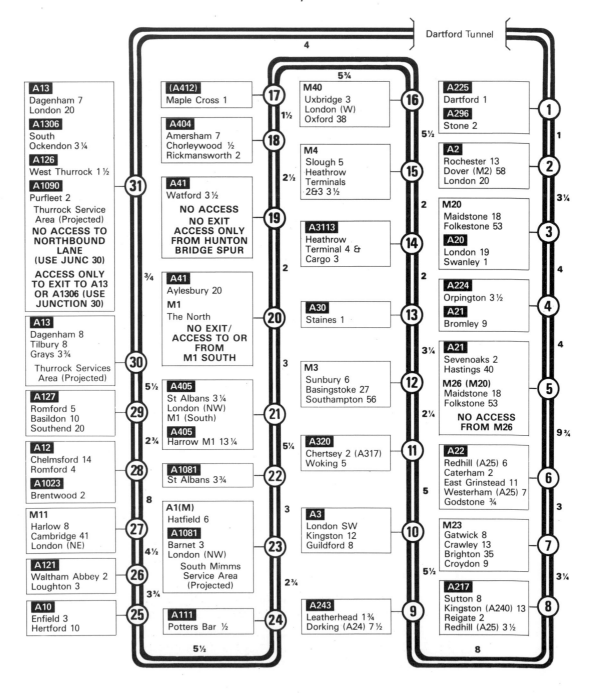

Exercise 5 Use the M25 motorway map on the previous page to answer the questions in this exercise.

1 How far is Southend from the M25 at Junction 29?

2 Which road do you take at Junction 15 on the M25 to get to Slough?

3 From which junction do you leave the M25 if you are going to:
(a) Watford   (b) Oxford   (c) Harlow?

4 Between which two junctions on the M25 is this photograph taken?

5 Which road do you take at junction 1 on the M25 to Dartford?

6 Between which 2 junctions is the Dartford Tunnel?

7 How far is it to Southampton from Junction 12 on the M25?

8 Which junction on the M25 would you use to get to Heathrow Terminal 4?

9 How far is it from Reigate to Sutton via the A217? Look at Junction 8 on the M25.

10 How far is it from Enfield to Hertford via the A10? Look at Junction 25 on the M25.

11 David is travelling from Guildford to Sutton. He takes the A3, then joins the M25 at junction 10 and leaves it at junction 8 for the A217. How far will he drive?

12 How far is it from Brighton to Oxford via the M23, M25 and M40?

13 Jane wants to drive from Hastings to Kingston to visit her boyfriend. She decides to go by the A21, M25 then A3. How far will she have to drive in taking this route?

14 Karen and Michael live in Staines and are going to visit friends in Amersham. They will take the A30, M25 then the A404. How far will they drive?

Distance, speed and time

Information ➷ **Speed** is a measure of the **distance** you travel in a unit of **time**. The units of time are usually 1 second or 1 hour. The units of distance are usually miles, kilometres or metres. The units for speed depend on the units used for distance and time.

Distance	Time	Speed
miles	hours	miles per hour (mph)
kilometres	hours	kilometres per hour (kph)
metres	seconds	metres per second (m/s)

If you travel a distance of 1 metre in 1 second your speed is 1 m/s.
If you travel a distance of 6 metres in 2 seconds your speed is 3 m/s.
If you travel a distance of d metres in t seconds your speed is s m/s.

There are 3 formulae which are used to solve speed problems.
In the formula **d** stands for **d**istance
　　　　　　s stands for **s**peed
　　　　　　t stands for **t**ime

$$d = st \qquad s = \frac{d}{t} \qquad t = \frac{d}{s}$$

Examples

Calculating time taken
Michael estimates that he can average 50 miles per hour (mph) on a journey to see his brother. His brother lives 190 miles away. How long will the journey take?

You need to calculate time (t)

$$t = \frac{d}{s} \qquad d = 190 \text{ miles} \qquad s = 50 \text{ mph}$$

time: $\frac{190}{50}$ = 3.8 hours   0.8 hours = 0.8 × 60 minutes
　　　　　　　　　　　　　　　　　　 = 48 minutes

Estimated driving time = 3 hours 48 minutes.

Michael will probably stop for a while once or twice on his journey, so add on another 30 minutes to make the estimate more realistic.

Estimated journey time is 3 hours 48 mins + 30 mins = 4 hours 18 minutes.

Calculating speed
A cyclist travels a distance of 54 kilometres in 2½ hours. What is her average speed?

You need to calculate speed (s).

$$s = \frac{d}{t} \qquad d = 54 \text{ km} \qquad t = 2.5 \text{ hours}$$

speed: $\frac{54}{2.5}$ = 21.6 kph

Her average speed is 21.6 kph.

Example

Calculating time

A runner can average a speed of 7½ miles per hour for 3 hours. Will he be able to finish a marathon which is 26 miles 385 yards in under 3 hours?

You need to calculate distance (*d*). $d = st$ $s = 7½$ mph $t = 3$ hours

Distance: $7.5 \times 3 = 22.5$ miles. This is less than 26 miles 385 yards.

He will not be able to finish the marathon in under 3 hours.

Exercise 6

Copy and complete the table below.

	Distance	Speed	Time
1	450 miles		9 hours
2	80 km		2½ hrs
3	500 m		80 seconds
4		40 mph	3 hrs
5		6 m/s	25 s
6		50 kpm	4½ hrs
7	90 miles	60 mph	
8	560 km	7 kph	
9	1600 m	4 m/s	
10	60 km		1½ hrs
11	240 km	32 kph	
12		36 mph	4¼ hrs

13 A jet aeroplane has an average cruising speed of 400 mph. How long will it take to fly 1000 miles?

14 A runner covers a distance of 6 miles in 48 minutes.
   (a) How many miles will she run in 1 minute?
   (b) How far will she run in 1 hour?
   (c) What is her speed in miles per hour?

15 Billy and Peter are trying to find out who can cycle the fastest. Billy manages to cycle 20 miles in 1¼ hours. Peter cycles 22½ miles in 1½ hours. Who can cycle the fastest and by how much?

16 On a 30 mile sponsored walk Mairi is noting how long it takes her to cover the distance. She takes 6½ hours to walk 26 miles.
   (a) What is her average speed?
   (b) How long will it take her to complete the walk?

Travelling in time

Information ↩   Remember that there are 60 minutes in 1 hour. You can do calculations when the time is given in hours and minutes by changing it to minutes. If you are dividing by time in the formula then multiply by 60, if you are multiplying by time in the formula then divide by 60 to change back into hours.

Examples   **Hours and minutes in calculations**

1   A lorry driver travels 54 miles in 1 hour 12 minutes. What is his average speed?

$s = \dfrac{d}{t}$ $d = 54$ miles $t = 1$ hr 12 mins which is 72 minutes

speed: $\dfrac{54}{72} \times 60 = 45$ mph ($\times 60$ to change back to hours)

2   A motorist drives at a steady speed of 90 kph for 2 hours 25 minutes. How far has he travelled?

$d = s \times t$ $s = 90$ kph $t = 2$ hrs 25 mins or 145 minutes
distance: $90 \times 145 \div 60 = 217.5$ kilometres ($\div 60$ to change to hours)

Exercise 7   1   An aircraft, cruising at an altitude of 35 000 feet and average speed of 390 miles per hour takes 1 hour 40 minutes to reach its destination. How far has it flown?

2   In his new car Oliver travelled 77 miles in 2 hours 12 minutes. What was his average speed?

3   When calculating the time for journeys Margaret reckons that she will average a speed of 45 mph. Use the mileage chart given earlier in the unit to find her expected journey time for the following trips (answer to the nearest minute):
(a) Edinburgh to Glasgow   (b) Glasgow to Leeds
(c) Leeds to Norwich   (d) Norwich to Hull
(e) Hull to Liverpool   (f) Liverpool to Glasgow.

4   A hill walker has an average walking speed of 4 kilometres per hour. He is planning a walk 21 km long. How long will the walk take? To allow for rests he adds on 10 minutes for every hour or part of an hour. What total time should he allow for the walk?

5   How long will it take Robert to run a half marathon of 13 miles if he reckons he can average 7 miles per hour? Give your answer to the nearest minute.

6   During a road test Kim drove a car at a steady 56 miles per hour for 2 hours 15 minutes. What was the total distance she covered?

7 Carol takes 4 hours 50 minutes to complete a 30 km sponsored walk. What is her average speed?

8 A ship sails for 4 hours 25 minutes at a speed of 15 knots (nautical miles per hour). How many nautical miles has it covered?

9 Derek is planning to walk from John O' Groats to Lands End, a distance of 874 miles as the crow flies, or 1103 miles by road.

In training he averages a speed of 3½ mph. He plans to walk for 8 hours a day, 6 days out of 7. How many days will the walk take?

10 The speed limit in built up areas is 30 mph. In a police speed trap a car is timed over a distance of ¼ mile. How many seconds will a car take to cover ¼ mile if it travels at 30 mph?

Here are the times for 4 cars.
(a) 28 seconds   **(b)** 32 seconds
(c) 30 seconds   **(d)** 25 seconds.
Which cars are breaking the speed limit?

11 The Flying Scotsman train leaves Aberdeen each week day at 0800 and arrives in London Kings Cross at 1508. The distance by rail is 524 miles. Calculate the average speed of the train.

12 A coach leaves York at 1130 for London. If the coach averages 45 mph when will it reach London?

13 A parachutist jumps from an aircraft at 10 000 feet. She free falls at 176 feet per second to 2500 feet when she releases her parachute. For how long does she free fall? Answer to the nearest second.

14 Bob has an appointment for a job interview at 11 am. He decides to drive to the interview, a distance of 36 miles.

He leaves at 9:50 am and covers the first 20 miles in 35 minutes. He then has a puncture which delays him for 15 minutes. What speed must he average over the last 16 miles to get to the interview in time?

15 The distance from Prestwick to New York is 5200 km. Calculate, to the nearest minute, the flight time for an aircraft which averages a speed of 680 kph.

Record breakers

Information ➭   In sporting events like racing and running, the fastest time for a certain distance is called the **record**. You can work out the average speed of the record holder if you know the record time and the distance of the race.

Example   **Record speeds**
The fastest greyhound ever recorded was a dog called The Shoe. He took 20.1 seconds to cover a distance of 410 yards on a straight track at Richmond, New South Wales, Australia on 25 April 1968.
What was his average speed in miles per hour?

410 yds in 20.1 seconds is average speed of 410 ÷ 20.1 yds per second
To change seconds to hours multiply by 60 × 60
To change yards to miles divide by 1760 (there are 1760 yds in a mile)
410 yds in 20.1 seconds = 410 ÷ 20.1 × 60 × 60 ÷ 1760 mph
Average speed = 41.7 mph

Exercise 8   For each of the records in questions **1** to **16** find the average speed in miles per hour or kilometres per hour.

1   The fastest racehorse recorded was a 4 year old named Big Racket who covered ¼ mile in 20.8 seconds at Mexico City on 5 February 1945.

2   The world speed skating record is held by Pavel Pegov (USSR) who covered 500 metres in 36.57 seconds at Medro, USSR on 26 March 1983.

3   The womens world swimming record was set on 21 July 1984 at Mission Viejo, California by Dara Torres (USA) who swam 50 metres in 25.61 seconds. She was 17 at the time.

Dara Torres

4   The world record for the mile is held by Steve Cram (GB) who ran the mile in 3 minutes 46.31 seconds. The record was set at Oslo Norway on 27th July 1985.

5   The fastest time for a K1 canoe to travel the 22.7 miles from Fort Augustus to Lochend was set in October 1975 by Andrew Samuel. He took 3 hours 33 minutes and 4 seconds.

6   The 100 mile British road cycling record was set by Ian Cammish in July 1983. He cycled the distance in 3 hours 31 minutes and 53 seconds.

7 The world record for the 100 metres is held by Calvin Smith (USA) who, on 3rd July 1983 ran 100 metres in 9.93 seconds. He was running at Colorado Springs, Colorado, USA at an altitude of 2195 metres.

8 The world record for the 800 metres is held by Sebastian Coe (GB) and stands at 1 minute 41.73 seconds. It was set at Florence, Italy on 10 June 1981.

9 The world dragracing record for piston engines is 5.48 seconds measured over 440 yards. It is held by Gary Beck (USA) and was set at Indianapolis in 1982.

10 The fastest recorded time for a marathon is 2 hours 7 minutes 11 seconds set by Carlos Lopes (Portugal) at Rotterdam, Netherlands on 20 April 1985. A marathon is 26 miles 385 yards long.

11 The Oaks flat horse race was won in 1982 in a record time by *Time Charter*. The horse completed the 1½ mile course in 2 minutes 34.21 seconds.

12 The Yorkshire three-peak fell race record is currently held by Hugh Symonds of Kendal AC. On 29th April 1984 he completed the 24 mile course in 2 hours 50 minutes 34 seconds.

13 The world record for the 30 km walk is held by Maurizio Damilono of Italy. On 6th April 1985 he completed the 30 km walk in 2 hours 6 minutes 7.3 seconds.

14 The fastest recorded speed for roller-skating was set by Guiseppe Cantarella of Italy on 28th September 1963. He covered a distance of 440 yards in 43.9 seconds.

15 The world record holder for tobogganing down the Cresta Run (in Switzerland) is Franco Gansser of Switzerland. On 16th February 1985 he completed 1 212.25 metres in 41.95 seconds.

16 The record time for cycling from Edinburgh to London, a distance of 610 km, is 18 hours 49 minutes 42 seconds. It is held by Cliff Smith and was set on 2nd November 1985.

Distance/time graphs

Information ➥ Journeys can be shown graphically by plotting the distance travelled against the time taken on **distance/time** graphs. Time is always plotted on the horizontal axis and distance on the vertical axis.

Example

Reading a distance/time graph
The graph shows the journey of a cyclist who travelled between 2 towns 40 miles apart. Using the graph work out
(a) The time the cyclist began the journey.
(b) The time at which he stopped.
(c) For how long he stopped.
(d) Between which times he was cycling the fastest.
(e) His average speed for the whole journey?

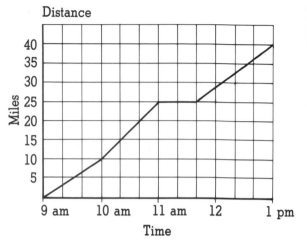

On the graph grid each hour is divided into 3. That is, each square represents 20 minutes. So

(a) He left at 9 am.
(b) He stopped for a rest at 11 am (the graph is a horizontal line starting here).
(c) He stopped for 40 minutes (until the graph slopes again).
(d) The steeper the graph, the faster the speed. So he was cycling fastest between 10 am and 11 am (between 10 and 25 miles).
(e) To work out his average speed use the formula $s = d \div t$ where $d = 40$ miles and $t = 4$ hours.
Speed $40 \div 4 = 10$ mph.

Exercise 9   1   The graph shows a car journey of 80 miles which Tricia and Mike took when they were in Yorkshire on holiday. They stopped twice to visit places of interest.

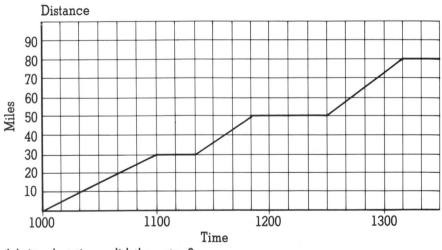

(a) At what times did they stop?
(b) How long did they spend on the second visit?
(c) How long in total did they spend travelling?

2   The graph shows Sarah's recent cycling trip to visit her Aunt.

(a) How far from Sarah does her Aunt live?
(b) When was Sarah cycling slowest?
(c) What was her average speed for the journey?

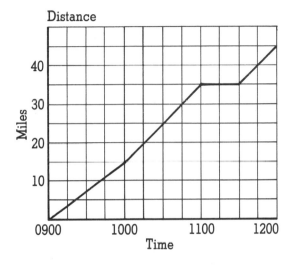

3   The graph shows Hazel's motorway journey. At one point she was held up by roadworks.

(a) At what time did Hazel meet the roadworks?
(b) How far did the roadworks stretch?
(c) How long did she stop?
(d) What was her average speed for the whole journey?

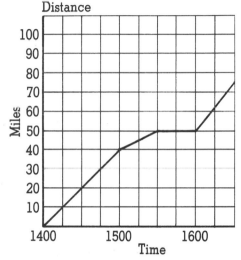

4   Dave recently travelled by car to the Lake District. Here is a graph of his journey.

(a) For how long did he stop?
(b) What was his fastest average speed during the journey?
(c) What was his overall average speed for the journey?
(d) At what time did he arrive at his destination?

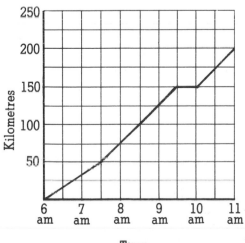

5   The graph shows how 3 people travelled from their homes in the same village to their work in the same town 10 miles away. Paul travelled by bus, Celia cycled and Jim travelled by car. The lines on the graph are labelled a, b, and c. Match one line to each person.

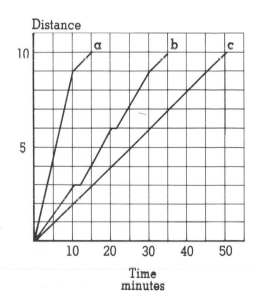

6   Ken and Jim went on a
cycling trip. The graph
shows both of their
journeys.

(a) For how long did
Ken stop?
(b) At what time did Jim
set off?
(c) At what time did
they both arrive?
(d) Calculate each of
their average
speeds.

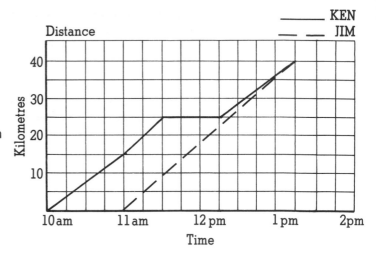

7   The graph shows how a
cyclist and a motorist
travelled between 2
towns 80 kilometres
apart.

(a) Who left first?
(b) Who arrived first?
(c) At what time (to the
nearest 5 minutes)
did the motorist pass
the cyclist?
(d) Calculate each of
their average
speeds.

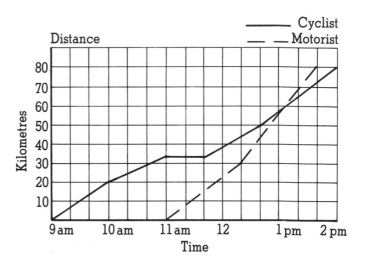

8   Peter and Derek live
40 miles apart. They
each decide to cycle to
visit friends in the other
town. Their journeys are
shown on the graph
below.
(a) When did Peter
leave?
(b) At what time (to the
nearest 5 minutes)
did they pass each
other?
(c) Who completed the
journey in the
shorter time?

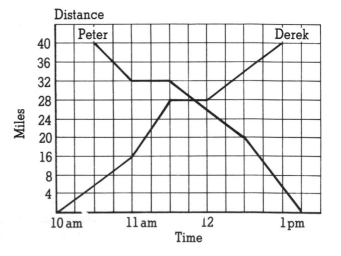

9 The graph shows how Mary and Gwen cycled between their 2 towns.
   (a) At what time did they pass each other?
   (b) How long did Gwen spend resting?
   (c) How long did Mary spend cycling?
   (d) Who covered the 50 km in the shorter time?
   (e) Who was the faster cyclist?

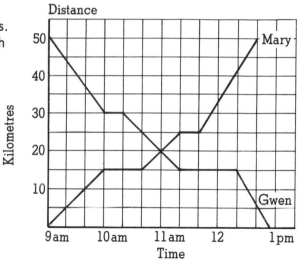

10 Liam is travelling to meet a client who lives 60 miles away. He left his office at 11 am and travelled the first 40 miles in 1 hour. He stopped for 30 minutes to have lunch and then completed the journey in half an hour. Show his journey on a distance time graph.

11 Steven is a sales rep. On Monday he leaves his house at 8 am and drives 30 miles to his first client. The journey takes him 1 hour. He is with the client for 30 minutes and then drives 50 miles for his second appointment. He arrives at 11 am and stays for 1 hour. He then spends ½ hour having lunch before driving 70 miles to his third client. The drive takes 2 hours. He is with the client for 45 minutes after which he drives 60 miles home, taking 1½ hours.

Show Steve's journeys for the whole day on a distance time graph.

12 Two friends live 60 kilometres apart. They arrange to meet at some point on the road between the towns. Keith leaves at 9:45 am and cycles for 1½ hours at an average speed of 16 kph. He then stops for 30 minutes for a rest. When he sets off again he cycles at an average speed of 18 kph.

Brian sets out at 10 am and covers the first 14 km in 1 hour. After a ½ hour rest he continues at an average speed of 16 kph.
   (a) Show both journeys on a distance time graph.
   (b) At what time do they meet?
   (c) Who had cycled further?

13 Sandra sets off to complete a 20 mile sponsored walk. In training she has averaged a speed of 4 mph. She plans to stop for a 15 minute rest every 2 hours. The walk starts at 10:30 am.
   (a) Show Sandra's walk on a distance time graph.
   Sandra's brother Graham intends to jog round the course and reckons to average a speed of 6 mph. He sets out at 11 am.
   (b) Show Graham's run on the same distance time graph.
   (c) At what time will Graham overtake Sandra?

Car insurance

Information ➾   In Great Britain it is illegal to drive a car without insurance. The cost of insurance depends on the age and driving experience of the owner, the district in which the car is kept, the type of car, the type of insurance etc.

In the following table the rates given are for 1 year and for a driver aged 30-49 years old who has held a full licence for at least 12 months. Younger drivers pay an additional percentage based on their age and driving experience.

3 means cover for Third Party, Fire and Theft only
c means Comprehensive cover

District	Insurance group											
	1		2		3		4		5		6	
	3	c	3	c	3	c	3	c	3	c	3	c
1	122	240	139	278	165	338	191	400	221	473	250	545
2	133	262	151	305	180	369	209	436	241	516	272	596
3	143	282	162	372	193	396	223	469	259	555	292	639
4	155	304	176	354	209	428	242	506	280	599	316	691
5	168	332	191	385	228	466	263	551	305	652	344	751
6	183	361	209	420	248	508	287	601	333	712	375	820

Rating Districts are determined by where you live. Rural areas like Borders Region have low ratings while city centres like Glasgow and London have high ratings.

The Insurance Group is determined by the type of car. Fast sports cars are in high groups while slower cars are in lower groups.

A **No claims bonus** is a discount given to drivers who do not make a claim on their insurance. The insurance rate is less the greater the number of years for which you do not claim. It is worked out as a percentage of the basic rate. For example:

1 Year 30%   2 years 40%   3 Years 50%   4 Years or more 60%

Examples

Calculating the cost

1 Dianne is thinking about buying an Escort 1.3L which is a group 3 car. She lives in Lothian which has a District Rating 2. She is 19 so an additional 50% is added. How much will it cost her to insure the car?

District Rating 2 Vehicle Group 3
From the table the basic rate for Comprehensive is £369
From the table the basic rate for Third Party Fire & Theft is £180
For Dianne
Comprehensive   £369+50% of £369 = £553.50
Third Party Fire & Theft   £180+50% of £180 = £270

2 Harry is 38 and owns a Maestro 1.6 HL which is in group 4. He has been driving for 2 years and has not claimed on his insurance. He lives in the country with district rating 2. How much will comprehensive insurance cost him?

District Rating 2, Group 4, so Basic Rate is £436
No claims bonus is 40%, so he pays 60% of £436 = £261.60

Exercise 10

Use the insurance table on the previous page to answer these questions.

1 Paul is 20 and has held a full driving licence for 18 months. He has bought a Metro 1.3 LE which is in insurance group 2. He lives in Tayside which has a district rating of 2. How much will comprehensive insurance cost if he pays an additional 50% because of his age?

2 Fiona is 42 years old and owns a Datsun Stanza 1.6 GL which is in group 5. She lives in the West End of Glasgow which has a district rating of 6. How much will comprehensive insurance cost her?

3 Mark is 17 and has just passed his driving test. He wants to buy a Mini. How much will third party, fire and theft insurance cost if his car is in group 2, his district rating is 3 and he has to pay an additional 80% because of his age and lack of experience.

4 Robert is an experienced driver aged 45 who has never claimed on his insurance. He owns an Escort 1.6 L which is in group 4 and lives in a town with district rating 2. How much will comprehensive insurance cost if Robert has a full no claims bonus?

5 Sheila has a Renault 5 GTL which is a group 3 car. She lives in Nottingham which has district rating 2. She is 27 and pays an additional 10%. Last year she insured her car against third party, fire and theft. This year she gets her first no claims bonus and decides to change to comprehensive insurance. How much more will she have to pay?

6 Give 3 reasons why it is more expensive to insure a car if it is kept in a city centre than if it is kept in a rural area.

Motoring costs

Information ⮕ Once you own a car it is not just insurance that costs extra money. You have to buy petrol and oil, pay for parts, and garage repairs, pay Road Tax and so on. It all adds up!

Example

Counting the cost!

In one week Angela worked out that she had driven 104 miles, in her Mini Metro from which she gets, on average, 40 miles to the gallon of petrol. She also bought a spare tyre for £13.50 and had the car serviced at her local garage. They charged her £80.00 plus 15% VAT. If petrol costs £1.80 per gallon how much did Angela spend on her car that week?

Distance = 104   Mileage per gallon = 40   Cost per gallon = £1.80

So Angela spent $\frac{104}{40} \times 1.80$ = £4.68 on petrol

Garage bill was £80.00+15% of £80

£80.00+£12 = £92.00

Spare tyre   = £13.50
Total for week   = £110.18

Exercise 11

1   Gerry owns a Capri and gets 36 miles per gallon of petrol. He is planning a motoring holiday and estimates that he will cover 1250 miles. How much will he spend on petrol if it costs £1.86 per gallon?

2   Dave has bought a new car and wants to find out how many miles it does to the gallon. He fills up the tank then drives 254 miles. To refill the tank he puts in 8.2 gallons of petrol. How many miles per gallon is Dave getting? (Answer to the nearest mile.)

3   Keith has kept a note of all his car expenses for the last year.
Road Tax £100   Insurance £228   Services £152
Tyres £72   Petrol £588   AA Membership £38.50
How much on average has Keith spent per week on his car?

4   Four friends are going away for a weekend. Assad has a Fiesta which averages 42 miles per gallon. Petrol costs £1.82 per gallon. If they drive a total of 236 miles, how much should each person pay Assad? Do you think a simple 4 way split is fair?

Youth club outing

Information ↩ Remember that Mary and Kirsten wanted to hire a minibus for the Youth Club outing.

They asked 2 local firms for their rates:

The total cost will depend on the mileage. Suppose they thought that they would travel 120 miles then the costs can be calculated:

BERTZ £16 + £0.04 × 120 = £20.80   MAVIS £ 9 + £0.11 × 120 = £22.20

So Bertz is cheapest for this distance. But is Bertz always the cheaper?

It would be easy to see which firm is the cheaper if a graph of distance against cost is drawn.

Make up a table for each company

BERTZ	
Cost (£)	Distance (miles)
16	0
16+2 = 18	50
16+4 = 20	100
16+6 = 22	150
16+8 = 24	200

MAVIS	
Cost (£)	Distance (miles)
9	0
9+ 5.50 = 14.50	50
9+11 = 20	100
9+16.50 = 25.50	150
9+22 = 31	200

On the same grid draw the graphs.

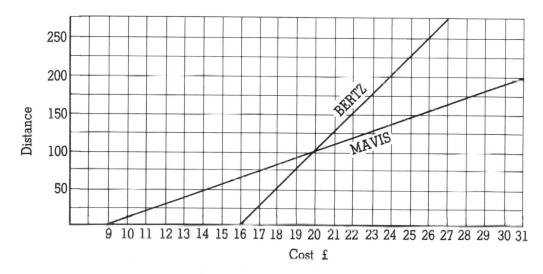

It is now easy to see that for any distance upto 100 miles Mavis is cheaper, but if you think you will travel more than 100 miles you should choose Bertz.

Exercise 12

1   A local car hire firm offers these 2 schemes for car rental.
    Scheme A   £18 per day plus 4p per mile
    Scheme B   £10 per day plus 8p per mile

    (a) Copy and complete the table below for each hire shceme.

Mileage	20	40	60	80	100	120	140
Scheme A							
Scheme B							

    (b) Plot a graph of cost against mileage showing both schemes on the same graph.

    (c) Between what mileages is scheme A best?

2   Scott wants to hire a caravanette for his family holiday. Two local firms offer him these rates.
    Firm A   £65 per week plus 4p per mile
    Firm B   £45 per week plus 6p per mile

    (a) Draw a table to show the cost of hiring a caravanette from each firm if he travels 200 miles, 400 miles, 600 miles etc.

    (b) Draw a graph to illustrate this information.

    (c) What advice should you give Scott about hiring the caravanette?

3   Roger is moving house and decides to hire a van. Here are 3 different hire rates he has been quoted.
    A   Fixed Price of £45 per day
    B   £20 per day plus 6p per mile
    C   £14 per day plus 8p per mile

    (a) Make up a table and find the cost of hiring a van using each scheme for different mileages.

    (b) Draw a graph of mileage against cost showing all 3 schemes of payment on the same graph.

    (c) Write down for which mileages each scheme is best.

4   Mr. Galloway wants to hire a car to go to a conference. He phones up 3 companies for quotes.

    BERT'S   car hire £9+8p per mile for distances over 100 miles.
    ECONOMY car hire £11+5p per mile for distances over 50 miles.
    DELUX   car hire No deposit, 12p per mile.

    (a) Make up a table to find the cost of hiring from each company.

    (b) Draw a graph of mileage against cost for each scheme.

    (c) Which scheme would be best if Mr. Galloway's conference was 135 miles away?

Investigation

Train times

It takes 45 minutes to travel between Edinburgh and Glasgow by train. There is a demand for a train to run in each direction every 30 minutes. Fifteen minutes must be allowed at the station for unloading and loading each train.

1   Draw up a daily passenger timetable from 07:00 to 19:00.

The graph below shows the timetable for 2 trains. The solid line shows the train which leaves Edinburgh at 07:00, arrives in Glasgow at 07:45, stands 15 minutes to unload and load passengers, leaves Glasgow at 08:00, arrives in Edinburgh at 08:45,....... The dotted line shows the timetable for the train which leaves Glasgow at 07:00.

2   (a) Copy the graph.

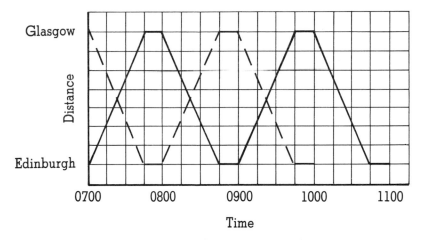

(b) Complete the graph for the train which leaves Glasgow at 07:00.

(c) Use another coloured pencil to draw on the graph the timetable for the train which leaves Edinburgh at 07:30.

(d) How many trains will you need to run the service?

3   To improve the service you are given 2 options:
(a) Newly designed trains which are faster and will complete the journey in 40 minutes — but you have to use the same number of these trains as you have at present.
(b) An extra train of the old type.

Which option will offer the greatest improvement in the service? In what ways will the service be improved?

Personal speed

What is your average walking speed?

Measure a distance of 50 metres in the playground. Use a stopwatch to measure the time it takes you to talk 50 metres at your normal pace. Use the formula $s = \dfrac{d}{t}$ to calculate your average walking speed.

Investigate whether your walking speed depends on the type of ground you are walking on — rough, hilly, grass, concrete etc.

15 Proportion

Do you remember? ⇨ In this unit it is assumed that you know how to:

1 use a rate in a problem

2 calculate a rate from given quantities

3 solve simple equations

4 substitute numbers for letters in equations.

Exercise 0 will help you check. Can you answer all the questions?

Exercise 0

Using a rate in a problem

1 Harry is a clerical assistant. It takes him 8 seconds to fold a letter, put it in an envelope and seal the envelope. He has 150 letters to send. How long will it take him to prepare all the letters?

2 Ramana types at a rate of 40 words per minute.
(a) How many words can she type in 15 minutes?
(b) How long will it take her to type a report of 20 000 words?

3 Claudette's car travels on average 32 miles per gallon of petrol.
(a) How far can she drive on 4 gallons?
(b) How many gallons of petrol will she need to travel 278 miles?

Calculating a rate from quantities

4 Find the cost of 1 video tape if 5 cost £22.50.

5 Joe can type 252 words in 6 minutes. What is his rate in words per minute?

6 Dave's car travelled 264 miles and used 8 gallons of petrol. What is its rate of fuel consumption in miles per gallon?

Simple Equations

Solve these equations.

7 $5x = 20$ 8 $8x = 64$ 9 $9y = 63$
10 $2.5v = 15$ 11 $5.4s = 27$ 12 $7.3k = 46.72$

Substitution

13 $s = 4k$ (a) Find s if $k=7$ (b) Find k if $s=64$
14 $d = 7.2l$ (a) Find d if $l=9$ (b) Find l if $d=46.8$

A sense of proportion

In Britain the Liberal Party and the SDP want to introduce a system of **proportional representation** to decide which party should govern the United Kingdom. They want to make sure that the total number of votes cast for a particular party is in direct proportion to the number of seats the party has in parliament.

If for example the Liberal Party get ¾ of the total votes cast in the country then they would get ¾ of the seats in Parliament.

In a small country there are 50 seats in parliament for 3 political parties, the Reds, the Greens and the Blues. Here are the total number of votes cast for each party. How many seats would each party be given with proportional representation? (Remember that all 50 seats have to be allocated.)

Party	Reds	Greens	Blues
Votes	34 750	96 260	76 060

This unit will help you solve problems involving proportion.

Direct proportion

Information ✑   Direct proportion is used in a number of everyday situations, for example rates of exchange, or time taken for a job. Two quantities are said to be in **direct proportion** if the increase (or decrease) in one is matched by an increase (or decrease) in the other. Direct proportion can be shown graphically.

Example   **Graphs of direct proportion**
John goes to the greengrocer and buys 5 lbs of potatoes for 80p. Sketch a graph and use it to find **(a)** the cost of 4 lbs of potatoes and also to find **(b)** what weight of potatoes you can get for 96p.

Step 1
Draw a set of axes with a suitable scale.

Step 2
Mark in the known quantities.
(0,0) will be one point since no potatoes will cost 0p.
(5,80) will be another point since 5 lbs of potatoes costs 80p.

Step 3
Join them up and extend the line.

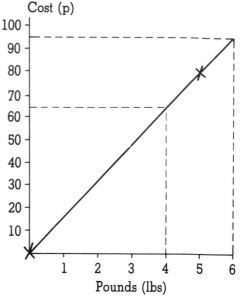

From the graph we can now read off the required values.
(a) 4 lbs of potatoes cost 64p.
(b) For 96p you can get 6 lbs of potatoes.

Exercise 1   In this exercise, draw a graph and use it to answer the questions. Remember that (0,0) will be one point to use.

1   8 km = 5 miles. Draw a graph.
Use it to estimate:

   (a) 3 miles in km
   (b) 6 km in miles
   (c) 8 miles in km
   (d) 10 km in miles

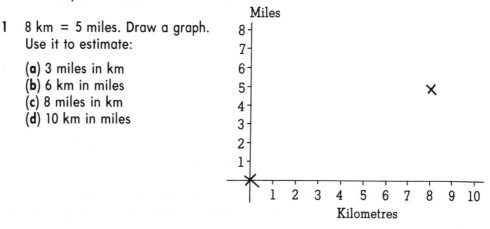

2   3 m of wood costs £7. Draw a
    graph. Use it to estimate:

    (a) The cost of 7 m
    (b) The cost of 2 m
    (c) The length of wood costing £12
    (d) The length of wood costing £5

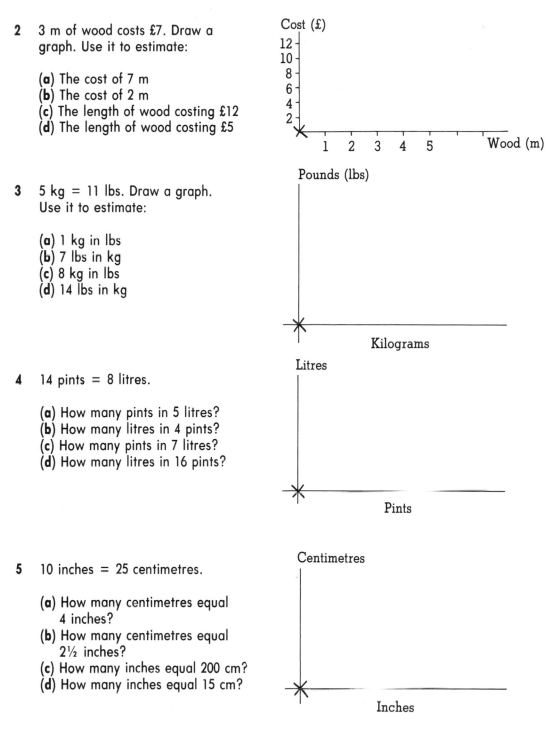

3   5 kg = 11 lbs. Draw a graph.
    Use it to estimate:

    (a) 1 kg in lbs
    (b) 7 lbs in kg
    (c) 8 kg in lbs
    (d) 14 lbs in kg

4   14 pints = 8 litres.

    (a) How many pints in 5 litres?
    (b) How many litres in 4 pints?
    (c) How many pints in 7 litres?
    (d) How many litres in 16 pints?

5   10 inches = 25 centimetres.

    (a) How many centimetres equal
        4 inches?
    (b) How many centimetres equal
        2½ inches?
    (c) How many inches equal 200 cm?
    (d) How many inches equal 15 cm?

6   4 m of cloth costs £10. Draw a graph and use it to find the cost of 9 m of cloth.
    What length of cloth can be bought for £12?

7   Buphinder pulls into a garage and fills his tank up with petrol. 34 litres cost him
    £16.32. How much would 20 litres have cost him? How many litres would he get for
    £10?

8   Mark's car travelled 200 miles on 12 gallons of petrol. How far will his car go on 10
    gallons. How many gallons will he need for a journey of 225 miles?

Finding other values

Information ➭   In the previous exercise you drew graphs to solve proportion problems. These problems can also be solved without using graphs.

Examples   **Direct proportion-ratio method**

1   In the example on page 261 John went to the greengrocers and bought 5 lbs of potatoes for 80p. You were asked to find how much 4 lbs of potatoes would cost. It can be done like this.

Weight (lbs)	Cost (p)
5	80
4	(less) $80 \times \dfrac{4}{5} = 64$

(The word *less* helps us to decide what fraction to multiply by.)

If the word is **less** the **smaller** number goes on the top line. If the word is **more** the **bigger** number goes on the top line.

2   4 tins of Bonzo dog food lasts Fifi 6 days. How long will 10 tins last?

Tins	Days
4	6
10	(more) $6 \times \dfrac{10}{4} = 15$

(More, so 10 is on the top line.)

Exercise 2

1   6 tins of Chumpy dog food last Rover 4 days. How long will 15 tins last.

2   Barry works a 30 hour week and gets £60. If he worked a 40 hour week (at the same rate) how much would he get?

3   Elizabeth is a bus driver. Her pay is £105 for a 35 hour week. How much would she get for a 25 hour week ?

4   It takes Bert 4 hours to paint an area of 60 m². How long will it take him to paint an area of 75 m²?

5   Ron can type a letter containing 300 words in 6 minutes. How long will it take him to type a letter containing 700 words?

6   A joiner can make 6 metres of fence in 45 minutes. How long will it take him to make a fence 32 m long?

7   Ian is a catering assistant who gets £84 for a 40 hour week. How much would he get for a 32 hour week?

8   Joe drives 400 kms in 5 hours. How far would he go in 6 hours. How long would it take him to travel 300 kms.

9   3 m of curtain material costs £3.60. Yaqub wants to buy 5 m. How much will it cost him. How much material could he buy for £10?

10   64 Francs can be exchanged for £8. How many Francs would I get for £25. How many pounds would be exchanged for 320 Francs.

Information ⇨   You must make sure that the two quantities have a rule connecting them in proportion before carrying out the sort of calculations you have been doing.

Examples   **Quantities not in proportion**

1   Henry VIII had 6 wives. How many wives had Henry IV?

There is no rule connecting the number of the king and the number of wives he had. In fact Henry IV had 2 wives.

2   An electrical shop is advertising video tapes for sale.

Number of tapes	1	5	10	20
Cost	£5	£23	£44	£85

Are the number of tapes and the cost in proportion?

No. If you pay £5 for 1 tape you would expect to pay £25 for 5 tapes.

Exercise 3   In these questions some of the quantities are in proportion, others are not. Answer the questions in which the quantities are in proportion. Answer *not in proportion* to those which are not.

1   An orchestra takes 54 minutes to play Beethoven's Sixth Symphony. How long will it take to play his fourth?

2   A television rental shop charges the following hire rates:

Screen size	14"	16"	18"	24"
Monthly rental	£7	£8	£9	£12

Is the hire charge in proportion to the screen size?
How much will a 22" television cost to hire?

3   A car hire company charges the following rates:

Miles	10	15	20	50	100
Charge(£)	4	5	6	15	35

How much will it cost to drive a hired car for 150 miles?

4   A post 4 metres high casts a shadow 2.5 metres long. A building is 18 metres high. What length of shadow will it cast at the same time?

Pie charts

Information ➱ **Pie charts** are used to display information. A circle is divided up into a number of different **sectors**. Each sector represents a different piece of information and the size of the sector is in direct proportion to the frequency of that information. The angle at the centre of the circle is used to find the size of each sector.

Example

Interpreting information from a pie chart
The pie chart below shows the results of a local council election. A total of 18 000 people voted.

We can get the following information from the pie chart.
(a) The Independants got the least votes.
(b) Labour won the election.
(c) Conservatives were second and Liberal/SDP third.

How many votes did Labour get?

The angle at the centre of the circle in the Labour sector is 162°.
360° represents 18 000 votes.

Angle	Votes
360°	18 000
162°	(less) $18\ 000 \times \dfrac{162}{360} = 8100$

Labour got 18 000 votes

Measuring angles
You may not be able to measure directly the angles in the pie charts, because the pie charts are not large enough for your protractor. However, if you take two pieces of paper and align the two edges along the two lines of the angle to be measured, you can use your protractor.

Exercise 4

1 Here are the results of a class survey on television viewing habits. Twenty students were asked which channel they watched most.

(a) Which was the most popular channel?
(b) Which was the least popular channel?
(c) How many students watched BBC2 most?

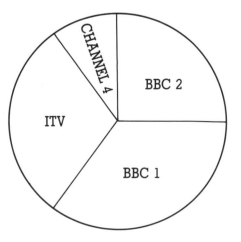

2 Two hundred people were asked which European country they would like to visit on holiday. The results are shown opposite.

(a) Which was the most popular country?
(b) How many people chose France?
(c) How many people chose Portugal?
(d) How many more people chose Denmark than Holland?

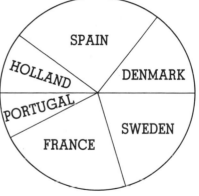

3 One hundred and fifty motorists were asked which make of car they would buy. The results are shown opposite.

Calculate the number of people who chose
(a) Ford
(b) Datsun
(c) Vauxhall
(d) Renault

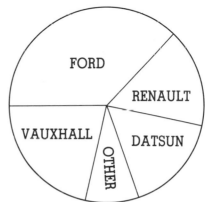

4 The pie chart opposite shows the age of the population of the United Kingdom for 1982. The total population was 54.8 million.

(a) How many people were under 16 years old?
(b) How many people were 65 years old or over?

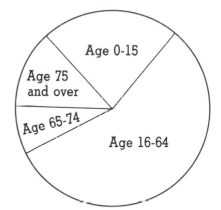

5 The pie chart shows the proportions of different types of trees which have been planted on an estate. Three hundred birch trees have been planted. How many spruce trees have been planted?

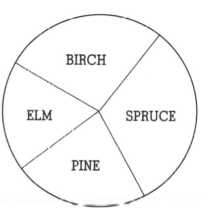

Example

Drawing a pie chart
The information below shows how Gordon spent his time last Wednesday. Show the information in a pie chart.

Sleep 8 hours, School 7 hours, Meals 2 hours, Sport 3 hours, Leisure 4 hours.

There are a total of 24 hours of activities in the list. Use direct proportion to work out the angle at the centre of each sector of a pie chart, and draw the pie chart.

Sleep

Hours	Angle
24	360°
8	(less) $360 \times \frac{8}{24} = 120°$

School

Hours	Angle
24	360°
7	(less) $360 \times \frac{7}{24} = 105°$

Meals

Hours	Angle
24	360°
2	(less) $360 \times \frac{2}{24} = 30°$

Sport

Hours	Angle
24	360°
3	(less) $360 \times \frac{3}{24} = 45°$

Leisure

Hours	Angle
24	360°
4	(less) $360 \times \frac{4}{24} = 60°$

Draw a circle of radius at least 5 cm. Draw 1 radius. Place the centre of your protractor on the centre of the circle and the zero line on the radius. Measure an angle of 120° and draw a second radius. Label the sector SLEEP. Move your protractor round and draw an angle of 105°. Label the sector SCHOOL. Continue until you have completed the pie chart. The last sector should measure 60°.

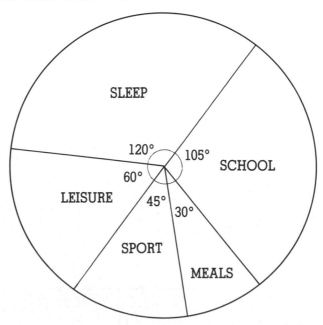

Exercise 5

1   Here are the results of a survey to find out which type of takeaway food is most popular.

Food	Chips	Chinese	Indian	Kebab	Burgers
Number	30	50	60	20	40

Show the information in a pie chart.

2   A group of people were asked to choose the make of computer they would like to own. The results are shown below. Show the information in a pie chart.

Computer	Acorn	Apple	Sinclair	Commodore	Atari
Number	150	46	115	74	95

3   The table below shows the share of the UK car market achieved by the 5 leading car manufacturers.

Austin/Rover	Ford	Vauxhall/GM	VW/Audi	Renault	Others
19%	26%	20%	6%	4%	25%

Show the information in a pie chart.

4   The table below shows the number of passengers, in millions, who used 6 of Britain's airports between November 1983 and November 1984.

Birmingham	Manchester	Glasgow	Stanstead	Heathrow	Gatwick
2.7	5.8	6.7	0.5	29	13.9

Show the information in a pie chart.

Inverse proportion

Information ⇨ **Inverse proportion** means that 2 quantities are connected in a different way. As one quantity increases the other decreases. For example the faster you travel the less time you take.

Examples **Inverse proportion**

1 Navinder completes a journey in 6 hours, travelling at 80 km/hr. Later a new section of motorway is opened and he can now travel at 100 km/hr. How long will Navinder take for the journey now?

Speed (km/hr)	Time (hours)	
80	6	
100	less	(less time because going faster)
	$6 \times \dfrac{80}{100} = 4.8$	(less so smaller number on top)
	4 hrs 48 mins	

2 Denise reads the instructions on how to cook a piece of meat in a microwave oven.

. . . for a 700 watt microwave cook for 4 mins . . .
(Time will be longer, proportionately, for microwaves with lower power ratings. Time will be shorter, proportionately, for microwaves with higher power ratings.

Denise's microwave has a 600 watt power rating. For how long should she cook the meat?

Power (Watts)	Time (mins)	
700	4	
600	more	(less power, more time)
	$4 \times \dfrac{700}{600} = 4.66$	(more so smaller number on top)
	4 mins 40 secs	

3 Paul has a coin box for his electricity. For 50p he can use his 2 kilowatt heater for 5 hours. He buys a new 2½ kilowatt heater. How long will his new heater run for 50p.

Power (kilowatts)	Time (hours)
2	5
2.5	less
	$5 \times \dfrac{2}{2.5} = 4$

Exercise 6

1 In example 1, how long would Navinder take travelling at 70 km/hr

2 In example 2, how long would the meat take to cook in an 800 watt oven

3 Joe's Cafe ordered enough food to last 20 people 6 days. How long would the food last if the average number of customers in a day was 25?

4 Bill runs his own Building business. He reckons that with 45 workers he can complete a job in 12 weeks. The contract states the job must be finished in 10 weeks. How many extra workers will he need to employ?

5 The Daily News ran a Bingo Competition. 6 people claimed a share of the first prize of £1000. Another 2 readers also put in a claim. How much would each person get?

6 Jeff Bowman wrote a book of 310 pages with an average of 250 words per page. One edition is printed using large type which can only fit an average of 200 words per page. How many pages will this edition contain?

7 Louise is a sales rep. who averages a speed of 70 mph on the motorway and reaches her destination in 4½ hours. On the way back she averages 60 mph. How much longer does she take on the return journey?

8 Lee is a farmer. During the Summer he hires 20 people to help him with his harvest. It takes the 20 people 7 days to take in the crops. How long would it have taken if he had hired 25 people?

Direct variation

Information ⟿   Direction variation is a similar idea to that of proportion. Look at these examples.

Example   **Graph of direct variation**
A piece of elasticated rope stretches by different amounts according to the weight of the object hanging from it. Sketch a graph of stretch against weight. What is the rule connecting them?

Weight (w kg)	Stretch (s cm)
1	3
2	6
3	9

A graph can be plotted using these values. It is a straight line.

Look back at the table. For each pair of values calculate s ÷ w.

Weight (w kg)	Stretch (s cm)	s ÷ w
1	3	3
2	6	3
3	9	3

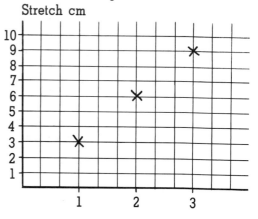

Stretch cm

weight (kg)

In each case s ÷ w = 3.

There is a rule connecting the weight hanging from the rope and the stretch of the rope. It is $\frac{\text{stretch}}{\text{weight}} = 3$ or $\frac{s}{w} = 3$ which can also be written as

stretch = 3 times the weight or s = 3w

2   A similar experiment was carried out on an elasticated rope of different diameter. The results are shown below.

Weight (w kg)	Stretch (s cm)	s ÷ w
1	2.5	2.5
2	5.0	2.5
3	7.5	2.5

The rule in this case is $\frac{s}{w} = 2.5$ or s = 2.5 w.

Example

Finding the rule

The Ace Van-hire Company charge according to the distance travelled. They charge £20 for a 4 mile trip. What is the rule connecting the distance and the charge?

Sketch the graph.

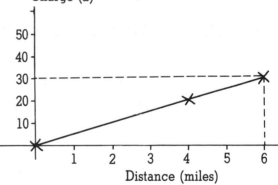

It is now easy to read off values for different trips. For example 6 miles would cost £30, 3 miles would cost £15.

$$\frac{20}{4} = 5 \qquad \frac{30}{6} = 5 \qquad \frac{15}{3} = 5$$

So the rule is

Cost ÷ distance = 5
or c = 5d

Exercise 7

For each of these questions
(a) sketch the graph
(b) write down the rule connecting the two quantities.

1   Taxi charges are £12 for a 4 mile trip. What would the charge be for a 3 mile trip, a 7 mile trip? Copy and complete the rule:
    charge   = 3 times the distance or c = ☐ × ☐

2   Potatoes cost 60p for 5 lbs. What would be the cost of 8 lbs? What would be the cost of 4 lbs? Copy and complete the rule:
    cost = ☐ times the   or c = ☐ × ☐

3   48 Francs can be exchanged for £3. How many pounds could you get for 53 Francs? Copy and complete the rule:
    pound = ☐

4   If it costs 60p to run an electric fire for 2 hours, how much would it cost to run the fire for 5 hours? What is the rule?

5   600 Pesetas can be exchanged for £3. How many pounds could I get for 800 Pesetas? How many pesetas could I get for £5?

6   A train travels 270 miles in 3 hours. How far will it travel in 4¼ hours? How long will it take to travel 400 miles?

7   In her car, Shabana can travel 240 miles on 5 gallons of petrol. How much petrol will she need to travel 384 miles?

8   A crate containing 24 bottles of beer costs £18. How much will 30 bottles of beer cost?

Information ↪ In each of the questions above you should have drawn a straight line graph through the origin (0,0). When two quantities produce a graph like this then they are said to **vary directly**. The symbol that is used is α. For example, $y \, \alpha \, x$ is read as y varies as x.

Example In the example on p.271 you found that $s = 3w$ The graph drawn was a straight line. This can be written as

$s \, \alpha \, w$ which reads s varies as w.

Exercise 8 In this exercise you are going to summarize the results of exercise 7. To do this you should copy and complete the table below, using your answers to exercise 7.

Question	Symbols	Rule	Graph
1	$c \, \alpha \, d$	$\dfrac{c}{d} = 3$ or $c = 3d$	
2	$c \, \alpha \, w$	$\dfrac{c}{w} = 12$ or $c = 12w$	
3	$p \, \alpha \, F$	$\dfrac{p}{F} =$	
4	c		

Information ↪ From the table you should see that if 2 quantities vary directly then one quantity is equal to a fixed number (called the **constant**) times the other quantity. The constant always has the same value. The two quantities are called **variables** because their values can vary.

Example **Calculating a value**

p varies as q. When $p = .24$, $q = 4$. Calculate p when $q = 7$.

$p \, \alpha \, q$ using 'k' as the constant term $p = kq$

Substitute for p and q $\quad 24 = k \times 4$
$\qquad\qquad\qquad\qquad 24 \div 4 = k$
$\qquad\qquad\qquad\qquad\quad k = 6$

Rewrite the formula $\qquad p = 6q$

Substitute for $q = 7$ $\qquad p = 6 \times 7$
$\qquad\qquad\qquad\qquad\quad p = 42$

Exercise 9

In each of the following questions
(a) find the value of the constant k
(b) write down the equation connecting the variables
(c) use it to calculate the new values.

1   12 Francs is equivalent to £2. How many Francs can you get for £36? Use $F \propto p$.

2   3 kg of potatoes cost 90p. How much for 7 kg? Use $c \propto w$.

3   4 bags of carrots cost 60p. How much for 9 bags? Use $c \propto b$.

4   y varies as x. When $x = 2$, $y = 12$. Calculate y when $x = 5$.

5   p varies as q. When $q = 4$, $p = 48$. Calculate p when $q = 7$.

6   m varies as n. When $n = 6$, $m = 30$. Calculate n when $m = 25$.

7   e varies as C. When $C = 17$, $e = 51$.
    Calculate (a) e when $C = 12$. (b) e when $C = 20$ (c) C when $e = 30$

8   Copy and complete this table. Write down the relationship connecting x and y.

x	2	4		8	
y	8		24		40

    Plot the points to check
    that it is a direct variation graph.

9   Copy and complete this table. Write down the relationship connecting electricity
    used and cost.

Electricity (units)	25	50	100		200
Cost (pence)		250		750	

10   This table shows the cost of admission to the cinema, depending on the size of
     group. Complete the table, sketch the graph and write down the relationship
     connecting the number of people and the cost of admission.

Number of people (n)	5		25	40	50
Cost in pounds (P)		25		100	

     How much would it cost for a group of
     (a) 6 people (b) 18 people (c) 30 people (d) 45 people
     (e) How many people would get in if £77.50 was paid?

11   The time taken to put up a fence is proportional to its length. It takes 4 hours to put
     up a 60 metre fence.
     (a) How long would it take to put up 90 m of fencing?
     (b) How much fencing could be put up in 8 hours?

A sense of proportion

Here are the results of the election.

Party	Reds	Greens	Blues
Votes	34 750	96 260	76 060

Total votes cast 34 750+96 260+76 060 = 207 070

207 070 must represent 50 seats in parliament. The parties seats are divided proportionally.

Reds

Votes	Seats
207 070	50
34 750	less

$$50 \times \frac{34\ 750}{207\ 070}$$

$$= 8.39$$

Greens

Votes	Seats
207 070	50
96 260	less

$$50 \times \frac{96\ 260}{207\ 070}$$

$$= 23.24$$

Blues

Votes	Seats
207 070	50
76 060	less

$$50 \times \frac{76\ 060}{207\ 070}$$

$$= 18.36$$

It is impossible to allocate a fraction of a seat in parliament so we must round all the answers to the nearest whole number. This gives

Party	Reds	Greens	Blues
Seats	8	23	18

There are 50 seats to be allocated but 8+23+18 = 49.

Who gets the last seat? Here are some possible answers:

(a) The party with most seats because they have a majority anyway.

(b) The party with fewest seats because they are not well represented.

(c) The party which came closest to getting another seat which in this case is the Reds.

Are there any other possibilities? Which do you think is fairest? Decide how you would allocate the last seat. Write down the reasons for your choice.

Investigation

Find out the total number of votes cast for the three major parties at the last General Election. How many seats would each party have under proportional representation? (There are 635 seats in total).

Find out how many seats each party actually has in Parliament. Which party would benefit most from proportional representation?

What are the advantages and disadvantages of proportional representation?

Exercise 10

1   Decide how to allocate seats for the following election results.

Party	Brown	Yellow	Orange
Votes	63 529	48 273	92 431

There are 40 seats to be distributed.

2   In an election in which 3 parties are represented the votes are:

Party A   Party B   Party C
34 298   16 825   60 946

There are 25 seats to be allocated. How many seats does each party get?

Explain how you decided to distribute the 25 seats.

3   Three friends do the football pools every week. They put in what they can afford and agree that if they win the money will be split in proportion to the amount each has contributed that week.

Last week they won £50 000 and can't agree how to split the money. Steve paid 80p, Wendy paid £1.00 and Kelvin paid 75p. How should it be split?

4   A school fund raising event earns £1000 to be divided as accurately as possible between the clubs. It is decided that the amount each club will receive will be proportional to its total membership. How much will each club receive?

Club	Football	Tennis	Cycling	Drama	Volleyball
Members	90	38	28	35	24

5   When communal repairs are done to a block of flats each person's share of the bill is proportional to their rateable value. A major roof repair costing £40 000 has just been completed. The rateable value for each flat is given below. How much must each person pay?

Barclay £815   Thornton £843   Mohammed £843
White   £815   Philip   £902   Murphy   £917

16 Space : the final unit

Do you remember?

In this unit it is assumed that you know how to:

1 compare, order and add positive and negative numbers
2 use a calculator
3 round numbers where appropriate.

Exercise 0 will help you check. Can you answer all the questions?

Exercise 0

Using integers

1 $-4+ \square = 6$ 2 $-3+ \square = 12$ 3 $\square +4 = 3$ 4 $\square +10= 4$

5 Put these numbers in order, starting with the largest 6, −3, 8, −9, 4

6 One winter's morning a thermometer rises from −4 degrees Centigrade to 1 degree. How much has it risen?

7 Your bank account is £20 overdrawn. You pay in £35. How much do you have in your account?

8 Lorna's bank account is overdrawn by £4. She pays in a cheque for £12. How much does she have in her account now?

9 50 BC means 50 years before our counting of years started. How many years are there between 50 BC and 1987?

10 How many years are there between 20 BC and the year you were born?

Using a calculator

Use a calculator to work out the answer to these sums. Round your answer to 1 decimal place.

11 1.6×2.6
12 3.4×2.6
13 5.8×8.6
14 5.4×4.5
15 6.9×2.4
16 2.4×8.1
17 $6.3 \div 1.2$
18 $8.6 \div 2.3$
19 $9.7 \div 3.4$
20 $8.9 \div 5.1$
21 $4.5 \div 7.3$
22 $1.2 \div 3.8$

Rounding numbers

Round these numbers to 2 decimal places.

23 3.141592
24 22.758
25 1.543
26 8.129
27 234.567
28 6.919
29 54.782
30 9.999

Star signs

Information ⟹ Here is Akpata looking at her horoscope. Her birthsign is Pisces. That means she was born between 19th February and 20th March.

PISCES

(Feb 20-Mar 20) People in authority should be more approachable and helpful now. Therefore, this is the perfect time for ideas and plans which lead to beneficial changes in your personal finances and at work.

ARIES

Her ruling planets are Neptune and Jupiter. She looks them up in a reference book and discovers that Neptune is 4.34×10^9 km from Earth and Jupiter is 6.28×10^8 km away. She doesn't understand this and asks her dad to explain the numbers. He is not sure either, so they try to figure it out. Can you imagine how far this really is?

In December 1986 the plane Voyager flew once around the world, non-stop. Look in a reference book and find out how far this is. How many times round the world would you have to go to cover a distance equivalent to going to Neptune or Jupiter?

What is your star sign? Do you know your ruling planet? How far away from Earth is it? What have your stars in store for you today?

After doing this unit you will be able to use and understand very large and very small numbers.

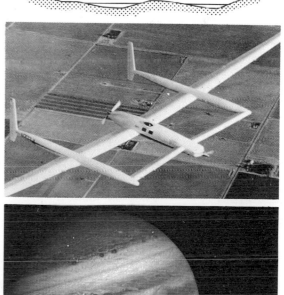

Jupiter from 28 million kilometres

Facts and figures

Information ⟹   All the planets in the table below are in our Solar System. They travel round the Sun on a constant path called an **orbit**. Each planet takes a different length of time to orbit the Sun. Some planets have a **moon** or moons which orbit round them.

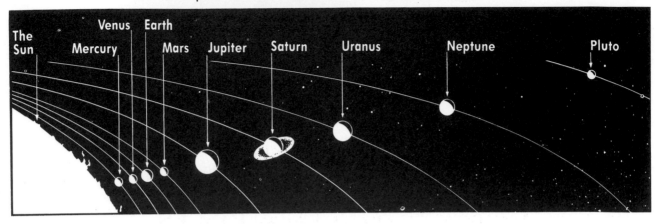

Here is some information on the Sun and the planets.

Planet	Diameter (km)	Average distance from sun (km)	Orbit time	Moon(s)
Sun	1 392 300	---	---	---
Mercury	4 880	58 000 000	88 days	0
Venus	12 100	108 000 000	224.7 days	0
Earth	12 756	150 000 000	365 days	1
Mars	6 790	228 000 000	687 days	2
Jupiter	142 800	778 000 000	11.9 yrs	16
Saturn	120 000	1 427 000 000	29.5 yrs	17
Uranus	49 500	2 869 000 000	84 yrs	5
Neptune	49 500	4 496 000 000	164.8 yrs	2
Pluto	6 400	5 940 000 000	247.7 yrs	1

Examples   **Using the table**

1   Find how long it takes for Jupiter to orbit the Sun.

Look down the column marked *Orbit time.*
Look along the row marked *Jupiter.*
The two meet at 11.9 yrs. That is the time it takes for Jupiter to orbit the Sun.

2   How much further from the Sun is Jupiter than Venus?

Look down the distance column. Find Jupiter. (778 000 000 km from Sun.)
Find Venus. (108 000 000 km from Sun.)
Difference is 778 000 000 − 108 000 000 = 670 000 000 km.

Exercise 1   Use the table to answer questions **1** to **10**.

1 Find the diameter of Pluto.

2 Find the distance from the Sun to Mars.

3 Find the time it takes Saturn to orbit the Sun.

4 How many moons does Saturn have?

5 How much further is Neptune than Jupiter from the Sun?

6 Find the time taken for Uranus to orbit the Sun.

7 Find the time taken by the Earth to orbit the Sun.

8 How much further from the Sun is Pluto than Mercury?

9 Find the time Venus takes to orbit the Sun.

10 How much closer to the Sun is Earth than Mars?

11 This picture of Earth was taken by a space shuttle as it approached the moon. Could you estimate how far away the shuttle was? That is, how far do you think it is from the Earth to the moon?

12 Find the diameter of Mars.

13 The Earth year (orbit) is 365 days. How long is a year on Mercury?

14 Find the diameter of Jupiter.

15 Find the difference in the diameters of Neptune and Pluto.

Standard form : large numbers

Information ⟹ The distances involved in space travel are enormous. Writing distances and doing calculations become laborious — you may have noticed this in Exercise 1! Instead of writing out such large numbers all the time a shorthand method is used. This is known as **standard form**.

The system uses **powers** of 10. Look at this table:

$$1\ 000\ 000 = 10^6$$
$$100\ 000 = 10^5 \quad \text{This is a quarter of the distance (in km) to the moon.}$$
$$10\ 000 = 10^4 \quad \text{This would be the capacity of a small football ground.}$$
$$1\ 000 = 10^3 \quad \text{Ann Boleyn was Queen for 1000 days (about 3 years).}$$
$$100 = 10^2 \quad \text{There are 100 years in a century.}$$
$$10 = 10^1$$
$$1 = 10^0$$

Can you see how the pattern develops?
Any large number can be written using this pattern.

Examples

Writing large numbers in standard form

1   Write 778 000 000 in standard form.

Write the number in 2 parts: 778 × 1 000 000
Make first part between 1 and 10:   7.78 × 100 000 000
Make second part a power of 10: 7.78×10^8

Notice that the number part is made to be between 1 and 10 and the powers of 10 adjusted to take account of this.

7.78×10^8 is an example of a number in standard form.

2   Write the distance from Pluto to the Sun in standard form.

From the table on p. 279, the distance is: 5 940 000 000 km
Write number in 2 parts: 594 × 10 000 000 km
Make first part between 1 and 10:   5.94 × 1 000 000 000 km
Make second part a power of 10: 5.94×10^9 km

3   Write the diameter of Earth in standard form.

From the table on p. 279, the diameter is: 12 756 km
Write number in 2 parts (first part
    between 1 and 10):   1.2756 × 10 000
Write second part as power of 10: 1.2756×10^4
Round to 2 decimal places: 1.28×10^4

Exercise 2   Write the numbers in questions **1** to **18** in standard form. Round to 2 decimal places if necessary.

1 27 000	**2** 14 000	**3** 5 000
4 300	**5** 20	**6** 453 000
7 123 000	**8** 67	**9** 93 000 000
10 4583	**11** 2 763 000	**12** 13 876 165
13 956 325	**14** 45 000 027	**15** 32 164
16 543 084	**17** 23 982 145	**18** 23 000 000 000

19 The moon is 384 000 km from the Earth. Write this in standard form.

20 Rewrite the table on page 281, using standard form where appropriate.

21 The Earth does not orbit the Sun in a circle, the path is shaped more like an ellipse. The closest that the Earth gets to the Sun is 147 097 000 km. Write this distance in standard form.

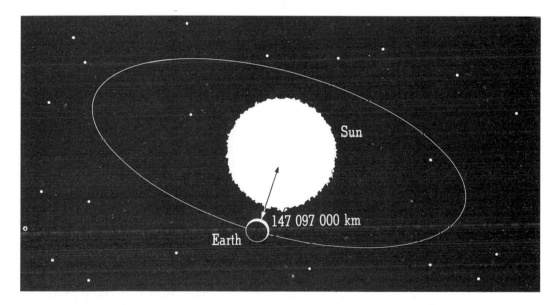

22 Light travels at 300 000 km per second. Write this speed in standard form.

23 The axis of the Earth is 10 000 km long. Write this length in standard form.

24 The Bristlecone pine trees of Nevada are the oldest trees known. The oldest of these is estimated to be 4900 years old. Write this age in standard form.

25 The number of different species of plant life is estimated at 350 000. Write this number in standard form.

26 The longest river in the world is the Nile in Africa. It is 6700 km long. Write this length in standard form.

27 The largest lake is the Caspian Sea, which covers an area of 423 400 square kilometres. Write this area in standard form.

283 Space: the final unit

Standard form : small numbers

Information ➯ As well as very large numbers, you may have to work with very small numbers. For example, the diameter of an atom is approximately 0.000 000 1 mm. Again, writing numbers like this would be laborious. We can extend standard form to take account of small numbers.

Look at the table again:

$$1\ 000\ 000 = 10^6$$
$$100\ 000 = 10^5$$
$$10\ 000 = 10^4$$
$$1\ 000 = 10^3$$
$$100 = 10^2$$
$$10 = 10^1$$
$$1 = 10^0$$

This continues to give:

$$0.1 = 10^{-1}$$
$$0.01 = 10^{-2}$$
$$0.001 = 10^{-3}$$
$$0.000\ 1 = 10^{-4}$$
$$0.000\ 01 = 10^{-5}$$
$$0.000\ 001 = 10^{-6}$$

and so on.

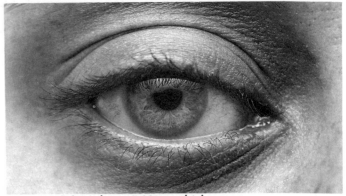

Your eyelid is about 0.001 m thick.

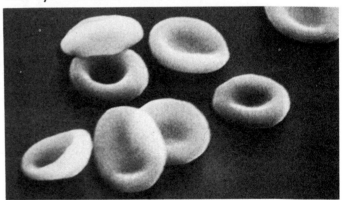

A blood cell is about 0.000 01 m long.

Examples

Small numbers in standard form

1 An orange contains 0.015 g of vitamin C.

Write this in standard form.

Write the number:	0.015
Split number into 2 parts:	15×0.001
Make first number between 1 and 10:	1.5×0.01
Make second number a power of 10:	$1.5×10^{-2}$ mm

2 The wavelength of visible light ranges from 0.000 04 cm for red light to 0.000 072 cm for blue light. Write these in standard form.

	Red light	Blue light
Write down number:	0.000 04	0.000 072
Split number into 2 parts:	4×0.000 01	72×0.000 001
Make first number between 1 and 10:	4×0.000 01	7.2×0.000 01
Make second number a power of 10:	$4×10^{-5}$ cm	$7.2×10^{-5}$ cm

Exercise 3   Write the numbers in questions **1** to **17** using standard form. Round to 2 decimal places where necessary.

1 0.000 56	**2** 0.000 002 3	**3** 0.006
4 0.000 000 086	**5** 0.005 64	**6** 0.0019
7 0.032	**8** 0.000 743	**9** 0.5
10 0.000 001 234	**11** 0.963	**12** 0.100 34
13 0.015	**14** 0.000 000 286	**15** 0.006 87

16   The smallest coin issued was the jawa, used in Nepal. The weight of the jawa was 0.014 g.

17   The photo shows a Bark Spider found in Nigeria. The smallest spider in the world is the patu marplesi. It is found in Samoa and is only 0.43 mm long.

Information   Numbers written in standard form can be converted to their original form simply by reversing the process you have used.

Examples   **Changing standard form to normal numbers**

1   Write out 1.36×10^6 in full.

$$\begin{aligned} 1.36 \times 10^6 &= 1.36 \times 1\,000\,000 \\ &= 136 \times 10\,000 \\ &= 1\,360\,000 \end{aligned}$$

2   Write out 8.3×10^{-4} in full.

$$\begin{aligned} 8.3 \times 10^{-4} &= 8.3 \times 0.000\,1 \\ &= 83 \times 0.000\,01 \\ &= 0.000\,83 \end{aligned}$$

Exercise 4   Write out the numbers in questions **1** to **12** in full:

1 1.56×10^2	**2** 3.2×10^3	**3** 6.3×10^5
4 2.34×10^{-1}	**5** 7.13×10^{-4}	**6** 9.12×10^{-2}
7 1.98×10^8	**8** 5.12×10^{-6}	**9** 7.2×10^{12}

10   The total amount of fish that could be caught in the oceans is 1.2×10^8 tonnes.

11   The potential arable land on Earth is 1.5×10^9 hectares of which about 4×10^9 acres is still unexploited.

12   An average cow gives about 2.7×10^3 litres of milk over a 10 month period. Why do you think the amount is measured over 10 months and not a year?

Calculations using standard form

Information ⟹ Standard form is used to make it easier to do calculations involving very large and very small numbers.

To **multiply** together two powers of 10 the powers are **added**.

Examples **Multiplying powers of ten**

1 $10^2 \times 10^3$

$(10 \times 10) \times (10 \times 10 \times 10)$
$10 \times 10 \times 10 \times 10 \times 10$
10^5

2 $10^2 \times 10^5$

$(10 \times 10) \times (10 \times 10 \times 10 \times 10 \times 10)$
$10 \times 10 \times 10 \times 10 \times 10 \times 10 \times 10$
10^7

Thus the product of two numbers expressed in standard form can be found by multiplying the two number parts together and adding the powers of 10.

Examples **Multiplying numbers in standard form**

Multiply these standard form numbers.

3 $(3.2 \times 10^2) \times (2 \times 10^3)$

$(3.2 \times 10^2) \times (2 \times 10^3)$
Collect the number parts together: $(3.2 \times 2) \times (10^2 \times 10^3)$
Multiply the number parts, add the powers: 6.4×10^5

4 $(4.1 \times 10^3) \times (3 \times 10^5)$

$(4.1 \times 10^3) \times (3 \times 10^5)$
Collect the number parts together: $(4.1 \times 3) \times (10^3 \times 10^5)$
Multiply the numbers, add the powers: 12.3×10^8
Make the number part between 1 and 10: $1.23 \times 10^1 \times 10^8$
Add the powers: 1.23×10^9

Exercise 5

1 $10^2 \times 10^4$ 2 $10^7 \times 10^5$ 3 $10^3 \times 10^8$ 4 $10^4 \times 10^6$

Write your answers to questions **5** and **16** in standard form. Round the number part of your answer to 2 decimal places if necessary. You may need a calculator for some of the multiplications.

5 $(1.2 \times 10^2) \times (2.3 \times 10^2)$ 6 $(3.2 \times 10^2) \times (1.4 \times 10^3)$

7 $(4.3 \times 10^4) \times (1.2 \times 10^2)$ 8 $(2.3 \times 10^5) \times (2.15 \times 10^1)$

9 $(1.5 \times 10^2) \times (3.5 \times 10^3)$ 10 $(4.7 \times 10^4) \times (6.2 \times 10^2)$

11 $(8.1 \times 10^8) \times (2.4 \times 10^2)$ 12 $(7.12 \times 10^1) \times (2.6 \times 10^7)$

13 $(4.32 \times 10^7) \times (6.31 \times 10^6)$ 14 $(5.16 \times 10^8) \times (2.81 \times 10^4)$

15 $(6.23 \times 10^6) \times (4.12 \times 10^2)$ 16 $(5.98 \times 10^7) \times (5.23 \times 10^7)$

17 A high speed machine makes 2×10^2 washers per hour. How many washers will be made in a week if the machine is used for 16 hours a day, 7 days a week? Express your answer in standard form.

18 A major electrical company employs 1.4×10^2 people each earning £8.2×10^3 per year. Express the annual wage bill in standard form.

Information ⟹ In **division** the process is similar to that of multiplication but the powers of ten are **subtracted**

Examples **Dividing powers of 10**

1 $10^5 \div 10^2$

2 $10^4 \div 10^2$

$(10 \times 10 \times 10 \times 10 \times 10) \div (10 \times 10)$
$10 \times 10 \times 10$
10^3

$(10 \times 10 \times 10 \times 10) \div (10 \times 10)$
10×10
10^2

Examples **Dividing numbers in standard form**

Divide these numbers in standard form. You may need a calculator for working out the number parts.

3 $(3.6 \times 10^4) \div (3 \times 10^1)$

Collect number parts together:
Divide numbers, subtract powers:

$(3.6 \times 10^4 \div (3 \times 10^1)$
$(3.6 \div 3) \times (10^4 \div 10^1)$
1.2×10^3

4 $(4.8 \times 10^7) \div (8 \times 10^3)$

Collect number parts together:
Divide numbers, subtract powers:
Make number part between 1 and 10:
Add powers:

$(4.8 \times 10^7) \div (8 \times 10^3)$
$(4.8 \div 8) \times (10^7 \div 10^3)$
0.6×10^4
$6 \times 10^{-1} \times 10^4$
6×10^3

Exercise 6

1 $10^6 \div 10^3$

2 $10^7 \div 10^2$

3 $10^4 \div 10^3$

4 $10^8 \div 10^3$

Write your answers to questions **5** to **18** in standard form. You may need a calculator for the number parts. Round the number part of your answer to 2 decimal places where necessary.

5 $(3.6 \times 10^4) \div (3 \times 10^1)$

6 $(4.8 \times 10^2) \div (4 \times 10^1)$

7 $(1.32 \times 10^3) \div (1.2 \times 10^2)$

8 $(6.3 \times 10^4) \div (2.1 \times 10^3)$

9 $(7.2 \times 10^6) \div (6.1 \times 10^2)$

10 $(9.9 \times 10^6) \div (3.3 \times 10^5)$

11 $(4.96 \times 10^7) \div (3.1 \times 10^3)$

12 $(6.7 \times 10^7) \div (3.1 \times 10^1)$

13 $(3.4 \times 10^8) \div (1.7 \times 10^4)$

14 $(5.6 \times 10^5) \div (1.12 \times 10^2)$

15 $(3.4 \times 10^9) \div (6.8 \times 10^2)$

16 $(1.3 \times 10^4) \div (3.9 \times 10^3)$

17 $(5.6 \times 10^8) \div (8.4 \times 10^6)$

18 $(2.6 \times 10^3) \div (7.5 \times 10^1)$

19 The three main engines in the space shuttle produce a thrust of 3×10^3 tonnes. How much is this per engine?

20 The space shuttle is covered with 3.1×10^4 heat-resistant tiles. The total surface area of the shuttle is 2×10^3 m². How many tiles per m² does the shuttle have?

Graphs with standard form

Information

Sometimes it is easier to make comparisons of large distances, speeds or numbers if they are shown on a **graph or chart**. When you draw a graph it is important to choose a suitable scale, to label the axes clearly, and to give it a title.

Example

Drawing a bar graph

Here is the table of facts and figures again. Draw a bar chart to show the average distances of the planets from the Sun. Write the scale you have chosen on the axis in standard form and label the axes.

Planet	Diameter (km)	Average distance from sun (km)	Orbit time	Moon(s)
Mercury	4 880	58 000 000	88 days	0
Venus	12 100	108 000 000	224.7 days	0
Earth	12 756	150 000 000	365 days	1
Mars	6 790	228 000 000	687 days	2
Jupiter	142 800	778 000 000	11.9 yrs	16
Saturn	120 000	1 427 000 000	29.5 yrs	17
Uranus	49 500	2 869 000 000	84 yrs	5
Neptune	49 500	4 496 000 000	164.8 yrs	2
Pluto	6 400	5 940 000 000	247.7 yrs	1

Convert each of the distances to standard form. Use the same power of 10 for each distance. Use 10^8.

Mercury is 58 000 000 km from the Sun. $58\,000\,000 = 0.58 \times 10^8$.
Venus is 108 000 000 km from the Sun. $108\,000\,000 = 1.08 \times 10^8$.
Earth is 150 000 000 km from the Sun. $150\,000\,000 = 1.5 \times 10^8$.
Mars is 228 000 000 km from the Sun. $228\,000\,000 = 2.28 \times 10^8$, and so on . . .

The number parts are then used to draw the bar chart.

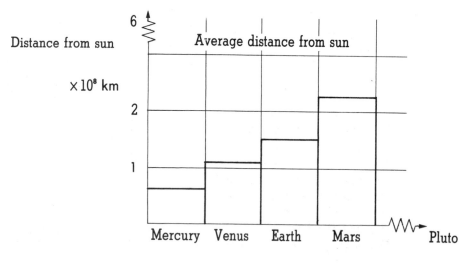

Exercise 7

1 Copy the bar chart and complete it for the other planets.

2 Draw another bar chart to compare the diameters of the different planets. Use it to write down the planets in order of size, starting with the largest.

3 Copy and complete this **pictograph** which shows the number of moons each planet has.

   ◗ = 1 Moon

Planet	Number of moons
Mercury	
Venus	
Earth	◗
Mars	◗ ◗

4 Draw a bar chart to show how long it takes for each of the planets to orbit the Sun. Put them in order, starting with the one which takes the least time to orbit.

Investigation

Standard form can be used in a lot of situations other than space travel. In any situation where you have to deal with very large or very small numbers you will find standard form helpful.

1 Here are some facts. Put the numbers in order of size, starting with the largest.
   A light year is 9.46×10^{15} m
   The mass of the moon is 7.23×10^{19} tons.
   The temperature of the Sun is 1.6×10^{7} degress Kelvin.
   The earliest recorded comet was in the 7th Century. The speeds of comets vary from 1125 km/hr to 2 000 000 km/hr. The closest approach to Earth was Lexell's comet on 1st July 1770 at a distance of 1.2×10^{6} km. The tail of the Great Comet (seen 1843) stretched for 330 million km.

2 Try to find out the information below. Express each answer in standard form, clearly stating the units you are using. Compare the answers for relative size by putting them in order from small to large.
   The average height of a skyscraper.
   The population of the world.
   The age of the planet Earth.
   The size of an atom.
   The size of an electron.
   The wavelengths of visible light.
   The size of a proton.
   The size of a virus.
   The size of a flea.
   The average height of a man.

A mixture of calculations

Information ➭ Multiplications and divisions are sometimes combined in one calculation. However you can do these by working out the number parts and the power part separately.

Example

Mixed calculation

Calculate $\dfrac{(2.4 \times 10^2) \times (1.2 \times 10^3)}{1.3 \times 10^2}$ giving your answer to 2 decimal places.

Separate the numbers and powers: $\dfrac{2.4 \times 1.2}{1.3} \times \dfrac{10^2 \times 10^3}{10^2}$

Multiply numbers: $\dfrac{2.88}{1.3} \times \dfrac{10^2 \times 10^3}{10^2}$

Add powers: $\dfrac{2.88}{1.3} \times \dfrac{10^5}{10^2}$

Divide numbers: $2.22 \times \dfrac{10^5}{10^2}$ to 2 dp

Subtract powers: 2.22×10^3

Exercise 8

Try these for size! You may need a calculator for the number parts. Simplify your answer to 2 decimal places where necessary.

1 $\dfrac{(2.7 \times 10^4) \times (1.7 \times 10^2)}{3.6 \times 10^3}$

2 $\dfrac{(3.1 \times 10^5) \times (1.2 \times 10^1)}{1.7 \times 10^2}$

3 $\dfrac{(7.6 \times 10^6) \times (4.7 \times 10^1)}{6.3 \times 10^3}$

4 $\dfrac{(8.1 \times 10^8) \times (4.5 \times 10^2)}{9.1 \times 10^7}$

5 $\dfrac{(4.3 \times 10^2) \times (3.7 \times 10^2)}{5.2 \times 10^2}$

6 $\dfrac{(5.6 \times 10^4) \times (2.7 \times 10^5)}{6.1 \times 10^6}$

7 $\dfrac{(7.5 \times 10^5) \times (2.4 \times 10^2)}{(3.2 \times 10^3) \times (1.2 \times 10^2)}$

8 $\dfrac{(6.3 \times 10^6) \times (4.3 \times 10^3)}{(2.4 \times 10^3) \times (5.3 \times 10^5)}$

9 $P = \dfrac{F}{a}$ is a formula used to calculate pressure.

Find P if $F = 6.3 \times 10^4$ and $a = 2.5 \times 10^2$.

10 $P = \dfrac{Fs}{t}$ is a formula which you used in Unit 3.

Find P if $F = 2.5 \times 10^3$, $s = 1.2 \times 10^2$ and $t = 6 \times 10^2$.

Star signs

Akpata wanted to find out how far away her ruling planets were. She looked them up and found that Neptune was 4.34×10^9 km away from Earth and Jupiter was 6.28×10^8 km away.

You should now be able to help Akpata.

Neptune: 4.34×10^9

$= 4.34 \times 1\ 000\ 000\ 000$
$= 4\ 340\ 000\ 000$ km away

Jupiter: 6.28×10^8

$= 6.28 \times 100\ 000\ 000$
$= 628\ 000\ 000$ km away

Can you think how far this actually is?

Compare it with the distance round the equator, which is approximately 4×10^4 km (40 000 km).

	Neptune	Jupiter
Distance from Earth / Distance round equator	$\dfrac{4.34 \times 10^9}{4.00 \times 10^4} = 1.08 \times 10^5$	$\dfrac{6.28 \times 10^8}{4.00 \times 10^4} = 1.57 \times 10^4$

That is, the distance from Earth to Neptune is the same as going round the world 108 000 times. The distance from Earth to Jupiter is the same as going round the world 15 700 times.

Did you find out about your ruling planet?

The distance to Jupiter is equal to going round the world 15 700 times

The distance to Neptune is equal to going round the world 108 000 times

Answers

Unit 1: Surveying

Exercise 0

1 10 mm, 100 cm, 1000 m 2 (a) 7 cm
(b) 2 cm (c) 3.5 cm (d) 4.7 cm
(e) 200 cm (f) 600 cm (g) 450 cm
(h) 375 cm 3 (a) 30 mm (b) 100 mm
(c) 85 mm (d) 42 mm (e) 500 mm
(f) 600 m (g) 1000 mm (h) 1100 mm
4 (a) 35° (b) 46° (c) 101° (d) 120°

Exercise 1

1 100 m, 50 m, 10 m, 40 m
6 (a) 37.2 cm (b) 82.8 cm
(c) 57.6 cm (d) 37.2 cm (e) 151 cm by
132 cm 7 2.8 m 8 1 km, 680 m 9 7.2 m

Exercise 2

2 25 cm 3 34 cm

Exercise 3

1 (a) 16 cm by 6 cm (b) 40 cm
2 67 m 3 4.5 km 4 16 cm by 11.2 cm
5 19 cm by 25 m 6 1 m by 1 m
7 1:40

Exercise 4

1 445 m 2 95 m 3 60 m 4 300 m
5 160 m 6 110 m 7 155 m 8 110 m
9 125 m

Exercise 5

1 (a) 70 m (b) 57 m 2 (a) 3.35 m
(b) 27° 3 8 m 4 m 4 1 m, 1.4 m,
2.2 m, 3 m

Exercise 6

1 68° 2 36° 3 x = 90°, y = 90°,
z = 51° 4 x = 98°, y = 126°,
z = 54° 5 65° 6 x = 51°, y = 78°

Exercise 7

1 67° 2 x = 127°, y = 53°,
z = 53° 3 96° 4 x = 74°, y = 74°
5 82° 6 x = 85°, y = 85°,
z = 85° 7 x = 100°, y = 80°,
z = 100° 8 x = 41°, y = 139°
9 x = 68°, y = 68°, z = 68°
10 x = 67°, y = 67°, z = 113°
11 a = 58°, b = 40°, c = 82°
12 a = 46°, b = 31°, PQ̂R = 77°
13 xŷ2 = 86°

Exercise 8

1 40° 2 48° 3 63° 4 52° 5 28°
6 76° 7 45° 8 50° 9 x = 78°,
y = 156° 10 x = 18°, y = 62°
11 70° 12 10°

Unit 2: Work and pay

Exercise 0

1 60 2 36 3 45 4 85 5 £1.80
6 £3.84 7 £3.40 8 £4.35 9 18 10 4
11 12 12 20 13 £2.20 14 £1.48
15 £2.60 16 £3.20 17 £1 18 £3.60
19 £57 20 £240 21 £4.38
22 £49.26 23 £36.40 24 £3.48
25 23h 59min 26 27h 32min
27 33h 44min 28 21h 54min
29 2h 9min 30 5h 9min
31 6h 55min 32 5h 45min

Exercise 1

1 £132.56, £41.58, £90.98 2 £839.00,
£221.59, £617.41 3 £242.00, £79.44,
£162.56.

Exercise 2

1 £61.25 2 £106.80 3 £94.72
4 £92.48 5 £76.00 6 £81.25
7 £63.90 8 £62.58 9 £45.36
10 £58.05 11 £1.98 12 £2.25
13 £2.45 14 £3.22 15 £2.84 16 £1.92

Exercise 3

1 £1.60, £1.80, £2.40 2 £2.25, £2.40,
£3.60 3 £2.55, £2.72, £3.06
4 £2.85, £3.42, £4.56 5 £3.15, £3.36,
£3.78, £5.04 6 £3.30, £3.52, £3.96,
£5.28 7 £3.75, £4.00, £4.50, £6.00
8 £3.90, £4.16, £4.68, £6.24
9 £85.50 10 £137.28

Exercise 4

1 £85.50 2 £104.50 3 £98.88 4 £99.22

Exercise 5

1 £168, £228 2 £142.50, £197.50
3 £168, £213 4 £56 5 £117.17
6 £175.50

Exercise 6

1 £58 2 £52 3 £45 4 £364 5 £176
6 £93 7 £62 8 £76
9 £4330, £57, £3660

Exercise 7

1 £8.84, £10.27, £19.11 2 £10.19,
£11.83, £22.02 3 £9.07, £10.53,
£19.60 4 £8.03, £9.33, £17.36
5 £10.15, £11.78, £21.93 6 £9.20,
£10.69, £19.89 7 £8.98, £10.42,
£19.40 8 £8.93, £10.37, £19.30
9 £8.66, £10.06, £18.72 10 £9.61,
£11.16, £20.77 11 £12.60

Exercise 8

1 £2425, £6575 2 £2525, £8975
3 £2525, £11 475 4 £2425, £10 050
5 £5382, £2227 6 £7454.20, £3689.20
7 £6552, £2187 8 £7488, £4245
9 £5820, £3555 11 £5441.80,
£2099.80 12 £7764, £3091

Exercise 9

1 £3795, £4705 2 Mr Davies: £4657,
£4558 Mrs Davies: £2425, £8232
3 £4332, £7987 4 Mr Chawla: £4223,
£9199 Mrs Chawla: £2425, £7175
5 Mr Edwards: £3795, £8205
Mrs Edwards: £2425, £9575

Exercise 10

1 379H 2 465H, 242H 3 433H 4 422H,
243H 5 379H 6 Single £2425-2429
7 Married £3420-3429 8 Single
£2450-2459 9 Married £4020-4029

Exercise 11

1 £2515, £754.50 2 £2796, £838.80
3 £2870, £861 4 £3057, £917.10
5 £3339, £1001.70 6 £2402, £720.60
7 £3795, £3205, £961.50 8 £2525,
£4363, £1308.90 9 £4180, £2840,
£852 10 £3127, £7589, £2276.70
11 £3454, £5346, £1603.80

Exercise 12

1 £4689.90 2 £5411.20 3 £6772.30
4 £6510.85 5 £6899.65

Exercise 14

1 More than £2600 2 £2000 or more
3 £2500 or more 4 £17 000

Unit 3: Symbolic Maths

Exercise 0

1 14 2 13 3 13 4 26 5 26 6 51 7 32
8 120 9 5 10 50 11 0.5 12 600
13 80 14 40 15 80 16 40 17 20
18 7 19 3.5 20 1 21 1 22 20 23 0.7
25 $3a$ 26 $5y$ 27 $8p$ 28 v 29 AC
30 RT 31 cd 32 d^2 33 y^2 34 s^2 35 $2t$
36 $5n^2$ 37 $10y$ 38 $15s$ 39 $48z$ 40 $12c^2$

Exercise 1

1 $60p$ $70c$ $65l$ 2 $76d$ $36t$ $54c$ 3 $39p$ $25o$
$18g$ 4 $37f$ $12t$ 5 $55h$ $56p$ $12m$ 6 $9a$
7 $7x$ 8 $6k$ 9 0 10 $20q$ 11 $12r+9s$
12 $5x+y$ 13 $5c+11d$ 14 $10f+8g$
15 $a+8b+7c$ 16 $7y+14z$
17 $6h+2j+7k$ 18 $2v+11w+4t$
19 $9s+7t$ 20 $21r+3s+t$ 21 $7ab$
22 $5xy$ 23 $5xy$ 24 $2ab+10pq$
25 $16x^2+4y^2$ 26 $9v+8v^2$ 27 $13x+5x^2$
28 $5c+d+6c^2$ 29 $5x^2+6y+2y^2+2x$

Exercise 2

1 $60p$ $80c$ $70l$ 2 $100d$ $40t$ $60c$ 3 $30p$
$12f$ $3t$ 4 $560s$ $92l$ $24t$ 5 $150G$ $200R$
$175E$
6 $2x+6$ 7 $3y+15$ 8 $12+6z$
9 $7a+7b$ 10 $15c+10d$ 11 $28f+14g$
12 $45x+36$ 13 $30y+20$
14 $6h+4j+8k$ 15 $32l+24p+16q$
16 $8a+12b+4c$ 17 $20+35m+20n$
18 $12x-30$ 19 $10c-35$
20 $24p-16q$ 21 $24g-30h$
22 $10c-15d-5e$ 23 $27p-54q-9r$
24 $12c-20d+4e$ 25 $35r+42s-14t$
26 $15u-10v-25w$
27 $4c-12d+8e-4t$
28 $6p+18q-30r-6s$
29 $8w-8x-48y+16z$

Exercise 3

1 $9a+31b$ 2 $28r+16s$ 3 $14x+29y$
4 $11g+42h$ 5 $22v+8w$ 6 $14s+3t$
7 $48x+10y$ 8 $25c+8d$ 9 $37v+4w$
10 $9c+7d$ 11 $19x+7y$ 12 $19p+5q$
13 $11x+22y+21z$ 14 $16a+27b+11c$
15 $14p+23q+14r$ 16 $11r+13s+10t$
17 $17w+16x+7y$ 18 $12x+8y+3z$
19 $8a+14b+10c$ 20 $9p+3q+19r$
21 $9c+2d+e$ 22 $18r+6s+2t$
23 $17r+6s+t$ 24 $8x+11y+4z$
25 $33j+14k+3l$ 26 $17r+27s+2t$
27 $8p+9q+13r$ 28 $11x+13y+3z$
29 $7c+d+9e$ 30 $13r+s+19t$
31 $22r+6s$ 32 $7x+13y$

Exercise 4

1 2 $(45p+35c+26l)$ 2 $2(45d+17t+9c)$
3 $5(112s+20l+7t)$ 4 $7(25G+42R+36E)$
5 $50(3p+s+r+2n)$ 6 $2(x+2)$
7 $3(a+5)$ 8 $5(l+2x)$ 9 $3(x+4y)$
10 $4(p+4q)$ 11 $7(y+z)$ 12 $5(x-3)$
13 $9(6-y)$ 14 6 $(p-5)$ 15 8 $(x-8y)$
16 $3(q-7r)$ 17 $10(r-s)$ 18 $4(2x+3)$
19 $3(4p+5)$ 20 $3(7+3r)$ 21 $7(2y+5z)$
22 $9(4s+3t)$ 23 $5(3v+w)$
24 $5(x+2y+3)$ 25 $3(2a+b+1)$
26 $4(3r+2s+5)$ 27 $3(3p+2q+4r)$
28 $2(5r+8s+3t)$ 29 $5(d+3e+5f)$
30 $2(2c-3d+5)$ 31 $3(5r+3s-7t)$
32 $5(10c-2d+7e)$ 33 $3(3p-6q-2)$
34 $6(3c-2d-5e)$ 35 $7(2g-4h-i)$
36 $x(5+y)$ 37 $p(q+7)$ 38 $c(d+4)$
39 $t(s-8)$ 40 $s(r-5)$ 41 $v(u-4)$
42 $p(g+r)$ 43 $d(c-e)$ 44 $x(y+z)$
45 $c(a+1)$ 46 $p(1-q)$ 47 $t(1-s)$
48 $3x(y+2z)$ 49 $2c(2+3d)$ 50 $5q(p+3r)$

Exercise 5

1 £3.98 2 £702 3 £161.50 4 £1610
5 £27.30 6 9 7 3 8 2 9 24 10 27
11 3 12 6 13 22 14 12 15 13
16 38 17 3 18 12 19 13 20 3 21 0
22 18 23 48 24 6 25 17 26 18
27 24 28 18 29 105 30 72 31 18
32 9 33 16 34 64 35 52 36 36
37 13 38 18 39 5 40 48 41 12 42 72
43 10 44 11 45 4

Exercise 6

1 200 2 296 3 24.5 4 328 5 962
6 1240 7 88 8 1235 9 8.58 10 141
11 1.73 12 10.5 13 322 14 296,
17.2 15 32.5 16 34 17 20 18 2.03
19 6.57 20 3770 21 40 000 22 603
23 49.6 cm 24 4.7 hr 25 1500
watts 26 5.26

Unit 4: Navigation

Exercise 0

1 Have a happy day 2 (2,1) (5,3)
(3,4) (1,1) (5,3) (3,3) (5,5) (2,1) (4,3)
3 24 4 3.8 5 5.4 6 0.72 7 0.69
8 37 m 9 45 m 10 4165 m 11 56.3 km
12 2.56 km 13 4.56 km

Exercise 1

1 600 300 2 200 100 3 400 500 4 300
600 5 300 300 6 400 000
7 Glasgow 8 Birmingham
9 London 10 Sheffield 11 Exeter
12 Leeds, York

Exercise 2

1 Blubberhouses 2 Bubwith
3 Wharram-le-street 4 Sledmere
5 3070 6 5050 7 5070 8 5010

Exercise 3

1 Harome Heads Farm 2 Sunley Hill
3 Boon Woods 4 Lowoods Farm
5 Cockpit Hall 6 Pasture House
7 Bowforth 8 Nawton Tower
9 Skiplam 10 Shaw Moor Farm
11 Fadmoor 12 Welbom Hall Sch
13 675 893 14 652 869 15 680 858
16 636 886 17 611 836 18 610 867
19 635 890 20 635 866 21 693 853
22 660 809 23 634 821 24 665 827
25 614 839 26 677 858 27 668 827
28 646 820 29 679 846 30 655 850

Exercise 4

1 9 km 2 3.25 km 3 4 km 4 950 m
5 14.3 km 6 380 m 7 31.5 km
8 11.2 km 9 1.2 km

Exercise 5

Answers are appropriate.
1 30 km 2 48 km 3 14 km 4 4 km
5 2 km 6 48 km 7 79 km 8 17 km
9 34 km 10 15 km 11 48 km
12 74 km

Exercise 6

1 W 2 NW 3 N 4 S 5 SW 6 NW
7 W 8 N 9 S

Exercise 7

(u) WSW (b) NW (c) SSE (d) NNW
(e) NNE (f) NNE (g) NE (h) NNE
(i) SSW

Exercise 8

1 166° 2 160° 3 136° 4 154°
5 178° 6 154° 7 292° 8 086° 9 271°
10 000°

Exercise 9

1 B 030° 15 km, C 340° 5 km, D 150°
10 km, E 160° 5 km, F 128° 15 km
12 Braddock 248° Conqueror 315°
Delta 054° Fox 095° Exeter 153°

Exercise 10

1 875 m, 213° 2 1 km, 160°
3 1.55 km, 003°

Exercise 11

1 4h 17 min 2 5h 20 min

Unit 5: Equations

Exercise 0

1 5 2 3 3 4 4 1 5 2 6 4 7 3 8 7
9 0 10 5 11 10 12 11 13 1 14 3
15 7 16 sub 5 17 add 3 18 div by
4 19 mult by 2 20 mult by 6 21 sub
4 22 add 10 23 div by 7 24 −8
25 +12 26 ÷9 27 ×7 28 −9
29 ÷7 30 ×12 31 3 32 4 33 6
34 13 35 12 36 22 37 5 38 4
39 5 40-42 straight lines

Exercise 1

1 29 2 47 3 7 4 37 5 3 6 6 7 7
8 13 9 8 10 9

Exercise 3

1 4 2 5 3 2 4 7 5 7 6 3 7 6 8 5
9 7 10 9 11 12 12 18

Exercise 4

1 3 2 2 3 5 4 8 5 10 6 6

Exercise 5

1 3 2 5 3 1 4 2 5 4 6 4 7 1
8 5 9 2 10 4 11 3 12 2 13 3 14 5
15 9 16 5 17 5 18 9 19 9 20 5 21 9

Exercise 6

1 2 2 2 3 2 4 6 5 2 6 6 7 5 8 4
9 3 10 6 11 4 12 8 13 4 14 6 15 3
16 9 17 6 18 3 19 6 20 12 21 2
22 3 23 4 24 10 25 7 26 4 27 9
28 3 29 4 30 1 31 3 32 1 33 2
34 ½ 35 ½ 36 ½ 37 2½ 38 1½
39 3½ 40 ¼ 41 2 42 1¼ 43 4½
44 ¾ 45 2½

Exercise 7

1 225, 9, 13 2 108, 6, 5 3 54, 8, 6
4 720, 120, 5 5 80, 5, 3 6 65, 11, 6,
7 7 48, 11, 6 8 188.4, 6.37, 3.18
9 10, 77 10 24, 16, 1 11 5 m
12 25 m 13 £7000, 5 years

Exercise 8

1 $x+y$ = 12, line 2 $x+y$ = 10, no
line 3 $x+y$=8, no line 4 $x+y$ = 6,
line 5 $x+y$ = 7, no line 6 $x+y$ =
11, line

Exercise 9

13 x = 5 14 x = 4 15 y = 2
16 $x+y$ = 3 17 x = $y+3$
18 x = $y+$l 19 $x+y$ = 7 20 y = 6

Exercise 10

16 parallel lines 23 $y=3x+1$
24 $y=2x-1$ 25 $y=7-2x$
26 $y = 4-⅔x$

Exercise 11

1 (3,2) 2 (4,2) 3 (1,2) 4 (3,1)
5 (2,5) 6 (4,2) 7 (5,5) 8 (6,2)
9 (2,4) 10 (1,5) 11 (3,7) 12 (1,4)

Exercise 12

1 speed $\leqslant$ 70 mph 2 age $\geqslant$ 18
3 age $\geqslant$ 16 4 weight $\leqslant$ 52 kg
5 4 < 7 6 18 > 14 7 5 < 7
8 3 < 5 9 11 > 10 10 100 > 95
11 960 < 1000 12 103 > 100 13 695
< 700 14 13 > 12 15 1 m > 1
foot 16 x > 4 17 x < 1 18 z > 6
19 c > 7 20 t > 7 21 y < 3

Exercise 13

1 Adult £4 Child £3 2 wholemeal 45p
white 42p

Unit 6: Pythagoras

Exercise 1

1 between 5 and 8 2 between 12 and
19 3 5<x<9 4 7<x<15 5 8<x<11
6 7<x<16

Exercise 2

1 4.47 2 6.40 3 8.06 4 7.81 5 9.43
6 9.85

Exercise 3

1 6.71 2 7.14 3 4 4 4.90 5 7.94
6 12.7

Exercise 4

1 8.49 2 9.85 3 10.2 4 8 5 3.32
6 1.36 7 17.0 8 22.2 9 32.9 10 19.4
11 5 12 9.43 13 0.7 14 16.5 15 42.0
16 13.2 17 14.9 18 52.2 19 5.66
20 8.49 21 7.07

Exercise 5

1 18.6 cm 2 8.54 cm 3 25.5 cm
4 12.2 cm 5 2.95 m 6 296 cm (inc
cross) 7 2.52 m 8 36.1 cm, 3.61 m
9 3.71 m 10 3 m 11 17.7 cm 12 132
km 13 29 cm 14 2.5 m 15 2.5 m

Exercise 6

1 8.94 2 4.58 3 3.61 4 5.59 5 11.2
6 6.98 7 14.1 m 8 17 cm

Exercise 7

1 10; 3,4,5; 2 2 39; 5,12,13; 3 3 12;
3,4,5; 4 4 32; 3,4,5; 8 5 130; 5,12,13;
10 6 60; 5,12,13; 5

Unit 7: Saving, Spending, Surviving

Exercise 0

1 £16 2 £17 3 £21 4 £43.65 5 £186
6 £443.80 7 £18.90 8 £198 9 20%
10 10% 11 25% 12 12.5% 13 10%
14 5.4% 15 70.4% 16 19.7% 17 £9
18 £40 19 £45

Exercise 1

1-9 Final balance £301.06 10 £58.64
11 overdrawn £6.10 12 £38.53
13 £24.22

Exercise 2

Final balance £543.90

Exercise 3

1 £45 2 £150 3 £2.50 4 £8.40 5 £3.75
6 £40.83 7 £3.25 8 £53.32 9 £50.31
10 £10 11 £47.60 12 £11.40 13 7%
14 6%

Exercise 4

1 £61.80 2 £66.56 3 £496.50
4 £287.19 5 £143.26 6 £135.52
7 £104.53

Exercise 5

1 £2.25 2 £62.75 3 £67.94 4 510.14
5 £290.17 7 £185.38

Exercise 6

1 £20.48, £91,52 2 £45.57, £120.26
3 £91.20, £1377.60 4 £48.83,
£128.94 5 £91.39, £840.04
6 £195.83, £1799.88 7 £35.52,
£40.24 8 £2227, £48 620

Exercise 7

1 £46.25 2 £20.32 3 £11.10 4 £25.50
5 £7.80 6 £56.40 7 £3.45 8 £42.08
9 £103.96 10 £22 11 £68.24 12 £98.40

Exercise 8

1 10% 2 25% 3 5% 4 20% 5 15%

Exercise 10

1 £0.70 P **2** £1.16 L **3** £17.40 P
4 £6.20 L **5** £85.50 P **6** £9.80 P
7 £0.40 P **8** £4.95 L **9** £0.31 P

Exercise 11

1 2.7% **2** 5.4% **3** 10.4% **4** 2.4%
5 55.3% **6** 2.5% **7** 0.7% **8** 6.8%
9 11.6% **10** Disks **11** Tool sets

Unit 8: D.I.Y.

Exercise 0

1 552 **2** 540 **3** 324 **4** 8.64 **5** 16.32
6 11.34 **7** 17.68 **8** 14.88 **9** 12.16
10 3 **11** 3.46 **12** 3.71 **13** 2.88
14 9.93 **15** 30.57 **16** 2.79 **17** 6.16
18 18.65 **19** 24 **20** 68 **21** 24 **22** 66
23 21 **24** 7 **25** 55 **26** 23 **27** 35
28 37.8 **29** 588 **30** 481 **31** 339 **32** 36.3
33 29.4 **34** 62.7 **35** 4.66 **36** 3.45
37 2.47 **38** 3670 **39** 28 600 **40** 243 000

Exercise 2

1 13 500 m² **2** 24 m² **3** 1600 cm²
4 600 mm² **5** 17.55 m² **6** cm²

Exercise 3

1 6 cm² **2** 30 cm² **3** 450 mm² **4**
5 cm² **5** 12.5 mm² **6** 26.65 cm²
7 70 m² **8** 455 mm²

Exercise 6

1 40 cm² **2** 120 m² **3** 175 m²
4 352.5 m² **5** 62.5 cm² **6** 140 m²

Exercise 7

1 (a) 154 cm² (b) 707 m²
(c) 254 mm² (d) 962 cm²
(e) 2290 mm² **2** 7.07 m² **3** 615 mm²
4 254 cm² **5** (a) 8 cm (b) 16 cm,
201 cm² (c) 6 mm, 113 mm²
(d) 12 mm, 452 mm² area×4

Exercise 8

1 346 cm **2** 9.42 m **4** (a) 8 cm,
25.1 cm (b) 16 cm, 50.2 cm
(c) 10 mm 62.8 mm (d) 20 mm,
126 mm (e) doubled

Exercise 9

1 31.14 m² **2** Yes **3** 11 m²

Exercise 10

1 (a) 2.5 l 1×2½ l £7.25 (b) 6 l
2×2½l+1×1 l £18.45
(c) 4.5 l 2×2½ l £14.50
(d) 7 l 3×2½ l £21.75
(e) 3 l 1×2½ l+1×1 l £11.20
2 £18.18 **3** £2.29

Exercise 11

1 8 **2** 8 **3** Kitchen 5 Bedroom 5
Lounge 7 Toilet 5 **4** Kitchen bedroom
bathroom **5** £5.50 £6.05 £9.90

Exercise 12

1 9ft of 2 m £42 **2** 14 ft of 3 m
£97.86 **3** 12 ft of 3 m £83.88 **4** 6 ft of
3 m £41.95 **5** 14 ft of 3 m £97.86 **6**
23 ft of 2 m £120.75 **7** 33 ft of 2 m
£173.25 **8** 35 ft of 2 m £183.75

Exercise 13

1 300 cm³ **2** 9 m³ **3** 0.35 m³
4 0.263 m³ **5** 11 664 cm³ **7** 2100 cm³

Exercise 14

1 480 cm³ **2** 2.925 m³ **3** 315 cm³
4 1700 cm³ **5** 13.2 m³ **6** 2261 cm³

Exercise 15

2 1 275 000 cm³ **3** (ii) (iv) **4** 1884 l
5 0.25 m³ **6** 30

Unit 9: Maths Abroad

Exercise 0

1 15.3 **2** 25.7 **3** 0.4 **4** 8.25 **5** £21
6 £300 **7** £187.50 **8** £170 **9** £3
10 £11 **11** £7.02 **12** £65 **13** 1680
4 1400 **15** 748 **16** 1170 **17** 3120
18 5060 **19** 30376.8 **20** 91621.8
21 3354.11 **22** 18.13 **23** 19.66
24 116.06 **25** 16.07 **26** 21.55
27 2.81 **28** £7.22 **29** £261.72
30 £3.83 **31** £4.24 **32** £1.51 **33** £12.06
34 £2.36 **35** £13.89 **36** £0.50
37 £12.84 **38** £1666.67

Exercise 1

1 £221 **2** £419
3 £347 **4** £218 **5** £195 **6** £399
7 £199 **8** £326 **9** £217 **10** No
snow! **11** 14 day holiday not
available **12** Gufo B&B only
13 San Lorenzo **14** 5, 12 Jan
15 29 Dec **17** £444 **18** £1506.05
19 £636 **20** £590.20

Exercise 2

1 £61.45 **2** £131.80 **3** £115.80
4 £177.90 **5** £438.70 **6** £211,
£272.45 (ii) £289, £420.80 (iii) £287,
£402.80 (iv) £195, £567.90 (v) £199,
£637.70

Exercise 3

1 Tenerife Aug 85 Oct 17 **2** Minorca
Jul/Aug 85 Oct 19 **3** Majorca
Jul/Aug 83 Oct 14 **4** Yugoslavia Jul
85 Jul 14 **5** Corfu Jul/Aug 90 Jul/Aug
18 **5** Portugal Jul/Aug 82 Oct 15
7 Malta Aug 86 Oct 19 **8** Madeira
Aug/Sept 76 Oct 17

Exercise 4

1 15.4 **2** 2.40 **3** 6.2 **4** 25.4 **5** 40 **6** 26,
24 **7** (c) No **8** 5.6 **9** 2.04 m **10** 71 kg

Exercise 5

1 £277.20 **2** £345.56 **3** £175.56
4 £347.15 **5** £390.85 **6** £406.06
7 £814.49

Exercise 7

1 800 gldr **2** 60900 esc **3** 969 fr **4** 156
800 lira **5** 1309 fr **6** 2597.50 kr
7 900.6 fr **8** 1076, 519.50 kr,
2034.90 kr

Exercise 8

2 £2.66 **3** £15.28 **4** £3.92 **5** £22.33
6 £12.40 **7** £15.54 **8** £28.41
9 3272.50 fr, 22.50 fr, £2.31 **10**
$357.50, £25.17

Unit 10: Sport

Exercise 0

1 25 2 45.6 3 40.6 4 20.2 5 16.8
6 3.8 7 17 9 114 kg 10 4.60%

Exercise 1

1 Mumtaz 2 Terry, Mumtaz, Harry, Linda 3 1 4 Peter, John, Alan, Keith, Assad, Waseem, Tom, Leslie 5 Fiona, Steve, Gary, Fiaz 6 Mumtaz, Linda 7 15 12 g = n−1 13 7 rounds 131 games 14 21 games

Exercise 2

1 Better goal difference 2 Sajid won more 4 Karen, Darren, Bob, Keith

Exercise 3

1 (a) decreasing (b) 1908, 1912, 1920, 1924, 1928, 1948, 1972, 1976, 1980 (c) 1932 (d) 20 years
2(a) decreasing (b) 1900 (c) 1906, 1920, 1948, 1964, 1972, 1976 (d) same as (c) plus 1980 (e) 1912

Exercise 4

1 2.33 2 13.6 3 81.6 4 15.2 5 27 000 6 4 hrs 9 mins 7 6 min 50 sec 8 763 9 3 hr 45 min

Exercise 5

1 89.2 2 28.25 3 47.4 4 79.3 5 55

Exercise 6

1 $\frac{1}{2}$ 2 $\frac{1}{2}$ 3 $\frac{1}{6}$ 4 $\frac{4}{13}$ 5 $\frac{1}{2}$ 6 $\frac{1}{50}$

Exercise 8

2 36, 1, $\frac{1}{36}$ 3 $\frac{1}{36}$ 4 $\frac{1}{6}$ 5 $\frac{1}{9}$ 6 $\frac{1}{7}$ 7 $\frac{1}{4}$ 8 $\frac{1}{16}$ (a) $\frac{1}{12}$ (b) $\frac{1}{12}$ (c) $\frac{1}{4}$ (d) $\frac{1}{4}$ 10 (a) $\frac{1}{26}$ (b) $\frac{2}{13}$ (c) $\frac{1}{8}$ (d) $\frac{1}{4}$

Unit 11: Right-angled triangles

Exercise 0

1 (a) 50 (b) 80 (c) 30 (d) 50 (e) 130 (f) 260 2 (a) 100 (b) 100 (c) 700 (d) 200 (e) 500 (f) 1500 3 (a) 2.7 (b) 6.4 (c) 54.8 (d) 148.6 4 (a) 60 (b) 50 (c) 90 (d) 100 (e) 40 (f) 50 5 (a) 240 (b) 670 (c) 2000 (d) 41 (e) 7.7 6 8 7 12 8 24 9 6 10 15 11 21 12 10 13 56 14 48 15 3 16 4 17 5 18 5 19 4 20 7 21 2 22 5 23 9

Exercise 2

1 0.637 2 1 3 2.45 4 15.056 5 25.4° 6 62.5°

Exercise 3

1 38.7° 2 41.2° 3 32° 4 37.9° 5 32.7° 6 33.7° 7 52.1° 8 51.3° 9 61.7° 10 65.4° 11 68.7° 12 54°

Exercise 4

1 38.7° 2 56.4° 3 47.2° 4 41.4° 5 44.4° 6 53.1° 7 33.7° 8 41.6° 9 33.7° 10 54.3° 11 42° 12 46.7° 13 41.2° 14 35.1° 15 44.4° 16 31.3° 17 39.8° 18 57.8°

Exercise 5

1 5.12 2 7.22 3 7.28 4 5.12 5 6.78 6 5.90 7 8.1 8 4.9 9 8.58 10 9.55 11 11.7 12 20.2 13 5.87 14 11.2 15 1.80 16 4.26 17 5.92 18 2.49 19 4.1 m 2.87 m 20 4.79 m

Exercise 6

1 7.93 2 29.1 3 12.9 4 16.4 5 9.09 6 18.9 7 9.00 8 8.17 9 9.04 10 26.0 11 10.2 12 17.2 13 13.0 14 17.2 15 29.2 16 33.2 17 12 18 18.8 19 4.31 m 20 30.2 m 21 7.14 km

Unit 12: Money and the home

Exercise 0

1 10 000 2 8000 3 4500 4 6650 5 23 400 6 14 300 7 £6300 8 £16 150 9 £43 225 10 £47 025 11 £32 937.50 12 £40 090 13 56 gals 14 £125 15 £182 16 £468, £9 17 Moira

Exercise 1

1 £32.31 2 £18.70 3 £30.58 4 £33.47 5 £21.63 6 £35.77

Exercise 2

1 (a) £6.25 (b) £41.62 (c) £2.91 (d) £5.24 2 (b) £4.31 (c) £33.30 3 £56.54 4 £94.85

Exercise 3

1 £5900 2 £2150 3 £6925

Exercise 4

1 £57.50 2 £51.75 3 £57.50 4 £74.75

Exercise 5

1 £188.46 2 £251.28 3 £366.45 4 £376.92 5 £418.28

Exercise 6

1 £588 2 £449.40 3 £324.24 4 £520.80 5 £373.80 6 £672 7 (a) £14 (b) £10.70 (c) £7.72 (d) £12.40 (e) £8.90 (f) £16 8 £61.50

Exercise 7

1 £0.75 2 £0.85 3 £1.15 4 £0.89, £170 000 5 £1.13, £50 000 6 £0.84, £800 000 7 £0.82 8 £1.15, £750,000

Exercise 8

1 £93.96 2 £56.70 3 £68.04 4 £77.76 5 £37.26 6 £69.66 7 £121.50 8 £113.40 9 £70.47 10 £72.17 11 £62.37 12 £99.87 13 £110.57 4 £51.48 15 £5.59

Exercise 9

1 £51.12, £57.96 2 £41.75, £47.33 3 £63.90, £72.45 4 £27.69, £31.40 5 £42.60, £48.30 6 £97.98, £111.09 7 £36.42, £41.30 8 £49.25, £55.83

Exercise 10

1 £7.20 2 £13.20 3 £12.32 4 £6.30 5 £10.60 6 £9.78

Unit 13: Trigonometry Problems

Exercise 0

1 48.6° 2 35.7° 3 60.9° 4 68°
5 31.4° 6 43.8° 7 11.3 8 18.4
9 6.78 10 19.6 11 4.13

Exercise 1

1 14.5° 2 48 m 3 1072 m 4 99.5 m
5 8.05° 6 2.32 m 7 18.3°

Exercise 2

1 85.6 m 2 74.6 m 3 17.8 m
4 23 m 5 31° 6 32.5° 7 66.4, 103,
36.6 8 82 m 9 28.2 m

Exercise 3

1 N 87.1 km E 134 km 2 E 24.5 km
S 52.6 km 3 S 426 km W 199 km
4 W 1.06 km N 2.38 km 5 51.3 km
6 214 km 7 56° 236° 8 067° 92.3 km

Exercise 4

1 33.4° 2 3.34 m 3 61.8° 4 2.52 m
5 55.4 m, 64 m 6 18.4° 7 7.19 m,
3.84 m 8 88.3 m 9 75 cm
10 (a) 1.3 m (b) 5.2 m²

Unit 14: Maths on the move

Exercise 0

1 1316 2 4964 3 84 240 4 2710.5
5 36 477.9 499.1 7 13 8 5.75 9 5.75
10 27.8 11 3.06 12 19.3 13 90
14 140 15 165 16 257 17 157
18 209 19 78 20 336 21 237 22 1:15
23 1:23 24 1:35 25 2:03 26 2:34
27 3:00 28 4:07 29 5:10 30 5:48 31 6
32 12 33 18 34 24 35 1 hr 30 min
36 36 37 1:42 38 1:54 39 08:00
40 11:30 41 12:00 42 14:45 43 20:20
44 23·55 45 14:15 46 21:45

Exercise 1

1 07:35 2 12:19 3 06:00, 7 hr 25 min
4 13:35, 2 hr 37 min 5 1310,
1 hr 2 min 6 08:00 is quickest
7 10:35, fewer stops 8 0800, 18 min
5 hr 42 min 9 11:24, 5 hr 42 min
10 13:10 11 07:47 12:39 4 hr 52 min
12 08:56, 55 min

Exercise 3

1 45 2 150 3 191 4 122 5 245
6 252 7 Hull 8 Dover 9 Janet by 115
miles

Exercise 5

1 20 miles 2 M4 3 (a) 19 (b) 16
(c) 27 4 15 and 16 5 A225 6 1,31
7 56 miles 8 14 9 6 miles 10 13 miles
11 29½ miles 12 109¾ miles
13 81.5 miles 14 24¾ miles

Exercise 6

1 50 mph 2 32 kph 3 6.25 m/s
4 120 miles 5 150 m 6 225 km
7 1½ hr 8 80 hr 9 400 s
10 40 kph 11 7.5 hr 12 153 miles
13 2½ hr 14 (a) 0.125
(b) 7.5 miles (c) 7.5 mph
15 Billy 1 mph 16 4 mph, 1 hour

Exercise 7

1 650 miles 2 35 mph 3 (a) 1 hr
(b) 4 hr 47 min (c) 3 hr 50 min
(d) 3 hr 24 min (e) 2 hr 48 min
(f) 4 hr 53 min 4 5 hr 15 min,
6 hr 15 min 5 1 hr 52 min
6 126 miles 7 6.21 kph 8 66.25
9 47 days 10 a, d 11 73.5 mph
12 16:08 13 43 sec 14 48 mph
15 7 hr 39 min

Exercise 8

1 43.3 mph 2 49.2 kph 3 7.03 kph
4 15.9 mph 5 6.4 mph 6 28.3 kph
7 36.3 kph 8 28.3 kph 9 164 mph
10 12.4 mph 11 35 mph
12 8.4 mph 13 14.3 kph
14 20.5 mph 15 104 kph 16 32.4 kph

Exercise 9

1 (d) 11:00, 11:50 (b) 40 min
(c) 2 hr 10 min 2 (a) 45 miles
(b) 0900-1000 (c) 15 mph
3 (a) 15:00 (b) 10 miles
(c) 30 min (d) 30 mph
4 (a) 30 min (b) 50 mph
(c) 40 kph (d) 11 am 5 (a) Jim,
(b) Paul, (c) Celia 6 (a) 45 min
(b) 11 am (c) 1:15 pm (d) Ken
12.3 mph, Jim 17.8 mph 7 (a) cyclist
(b) motorist (c) 1:10 pm (d) 16 kph,
30 kph 8 (a) 10:30 am
(b) 11:50 am (c) Peter
9 (a) 11 am (b) 80 min
(c) 2 h 40 min (d) Mary (e) Gwen
12 (b) 12:15 (c) Keith 13 Midday

Exercise 10

1 £457.50 2 £712 3 £291.60
4 £174.40 5 £86.13

Exercise 11

1 £64.58 2 £31 3 £22.67 4 2.56

Exercise 12

1 (c) 200 miles and over (2) A over
1000 miles, B under 1000 miles
3 A over 416 miles, B 300-416 miles,
C under 300 miles 4 (c) Delux

Unit 15: Proportion

Exercise 0

1 200 min 2 600, 8 hr 20 min
3 128, 9 4 £4.50 5 42 6 32 mpg
7 4 8 8 9 7 10 6 11 5 11 5 12 6.4
13 28, 16 14 64.8, 0.65

Exercise 1

1 (a) 4.8 (b) 3.75 (c) 12.8 (d) 6.25
2 (a) £16.30 (b) £4.70 (c) 5 m
(d) 2 m 3 (a) 2.2 (b) 3.2 (c) 3.6
(d) 31 4 (a) 8.75 (b) 2.3 (c) 86
(d) 9 5 £22.50 4.8 m 6 £9.60 20 l
7 167 miles, 13.5 gals

Exercise 2

1 10 days 2 80 3 £75 4 5 hrs
5 16 min· 6 4 hr 7 £67.20
8 480 km, 3¼ hr 9 £6, 8.33 m
10 200, £40

Exercise 3

2 Yes, £11 4 11.25 m

Exercise 4

1 (a) BBC1 (b) C4 2 (a) Spain (b) 45
(c) 15 (d) 10 3 (a) 50 (b) 25
(c) 35 (d) 30 4 (a) 11.9 million
(b) 8.2 million 5 400

Exercise 5

1 Chips 54° Chinese 126° Indian 80°
Kebab 36° 2 Acorn 113° Apple 35°
Sinclair 86° Commodore 56° Atari
71° 3 A/R 68° Ford 94° V/GM 72°
VW/Audi 22° Renault 14° Others
90° 4 Bir 17° Man 36° Gla 41° Sta 3°
Hea 178° Gat 85°

Exercise 6

1 6 hr 51 mins 2 3.5 min
3 4.8 days 4 9 5 £125 6 £388
pages 7 5½ hr 8 5.6 days

Exercise 9

1 $F = 6p$, 216 2 $c = 0.3w$, £2.10
3 $c = 15p$, £1.35 4 $y = 6x$, 30
5 $p = 12 q$, 84 6 $m = 5n$, 5
7 $e = 3C$, (a) 36 (b) 60 (c) 10
8 $v = 4x$ 9 $p = 5u$ 10 $p = 2.5n$,
(a) £15 (b) £27 (c) £75
(d) £112.50 (e) 31 11 (a) 6 hr
(b) 120 m

Unit 16: Space: The final unit

Exercise 0

1 10 2 15 3 −1 4 −6 5 8, 6, 4,
−3, −9 6 5° 7 £15
8 £8 92037 years 11 4.2 12 8.8
13 49.9 14 24.3 15 16.6 16 19.4
17 5.3 18 19.8 19 2.9 20 1.75 21 0.6
22 0.3 23 3.14 24 22.76 25 1.54
26 8.13 27 234.57 28 6.92 29 54.78
30 10.00

Exercise 1

1 6400 km 2 228 000 000 km
3 29.5 yrs 4 17 5 3 718 000 000
6 84 yrs 7 365 days
8 5 882 000 000 9 224.7 days
10 78 000 000 km

Exercise 2

1 2.7×10^4 2 4×10^4 3 5×10^3
4 3×10^2 5 2×10^1 6 4.53×10^5
7 1.23×10^5 8 6.7×10^1 9 9.3×10^7
10 4.58×10^3 11 2.76×10^6
12 1.39×10^7 13 9.56×10^5
14 4.5×10^7 15 3.22×10^4
16 5.43×10^5 17 2.40×10^7
18 2.3×10^{10} 19 3.84×10^5
21 1.47×10^8 22 3×10^5 23 1×10^4
24 4.9×10^3 25 3.5×10^5
26 6.7×10^3 27 4.23×10^5

Exercise 3

1 5.6×10^{-4} 2 2.3×10^{-6} 3 6×10^{-3}
4 8.6×10^{-8} 5 5.64×10^{-3}
6 1.9×10^{-3} 7 3.2×10^{-2}
8 7.43×10^{-4} 9 5×10^{-1}
10 1.23×10^{-6} 11 9.63×10^{-1}
12 1×10^{-2} 13 1.5×10^{-2}
14 2.86×10^{-7} 15 6.87×10^{-3}
16 1.4×10^{-2} 17 4.3×10^{-1}

Exercise 4

1 156 2 3200 3 630 000 4 0.234
5 0.000 713 6 0.0912 7 198 000 000
8 0.00 005 12 9 7 200 000 000 000
10 12 000 000 11 1 500 000 000,
4 000 000 000 12 2700

Exercise 5

1 10^6 2 10^{12} 3 10^{11} 4 10^{10} 5 2.76×10^4
6 4.48×10^5 7 5.16×10^6 8 4.95×10^6
9 5.25×10^5 10 2.91×10^7
11 1.94×10^{11} 12 1.85×10^9
13 2.73×10^{13} 14 1.45×10^{12}
15 2.57×10^9 16 3.13×10^5
17 2.24×10^4 18 £1.45×10^6

Exercise 6

1 10^3 2 10^5 3 10^1 4 10^5 5 1.2×10^3
6 1.2×10^1 7 1.1×10^1 8 3×10^1
9 1.18×10^4 10 3×10^1 11 1.6×10^4
12 2.16×10^6 13 2×10^4
14 5×10^3 15 5×10^4 16 3.33×10^0
17 6.67×10^1 18 3.47×10^1
19 1×10^3 20 1.55×10^1

Exercise 8

1 1.28×10^3 2 2.19×10^4 3 5.67×10^4
4 4.00×10^3 5 3.06×10^2 6 2.48×10^3
7 4.69×10^2 8 2.13×10^1
9 2.52×10^2 10 5×10^2

Index